Global Warming :
Myth or Reality

"As long as materialism is the new religion I can see no way people will reduce their consumption"

—Salik Rafiq, Blackburn

GLOBAL WARMING :
MYTH OR REALITY

Dr. ALURI JACOB SOLOMON RAJU
Associate Professor
Department of Environmental Sciences
Andhra University
Visakhapatnam 530003

MD PUBLICATIONS PVT LTD
NEW DELHI

Published by :

MD Publications Pvt Ltd
'MD House', 11, Darya Ganj
New Delhi - 110 002
Phone : +91-11-41563325 (Customer Service)
+91-11-41562846 (Editorial)
E-mail (orders): contact@mdppl.com
Website: www.mdppl.com

ISBN: 978-81-7533-143-3

Published and Printed by Mrs. Rajni Gupta on behalf of **MD Publications Pvt Ltd** at Times Press, 910, Jatwara Street, Darya Ganj, New Delhi.

PREFACE

The importance of environment is realized in recent times in view of the global warming and climate change. The greenhouse gases are the principal culprits for warming the globe and in effect causing the change in climate. The changing climate has been showing negative effects on most of the species. Some species have become extinct and some others are on the way to extinction. The still surviving species are also experiencing difficulty in reproduction and hence, population collapse and the attendant ecosystem collapse. The humanity is also facing food insecurity, nutritional insecurity and livelihood insecurity. Keeping these realities in view, this book has been written based on the information that has been published and displayed in different websites on the internet. I do not claim any part of the content presented in this book as my own unless otherwise specified. Further, the information thus collected has been analyzed and the same has been presented in a comprehensive manner for clarity.

This subject is receiving greater attention in recent times due to questionability of survival of humanity and biological organisms. This subject is very important for the students, researchers and academicians of biology, environmental science and all others who are simply human beings. This book is an important outcome for the realization of present state of human life and for taking corrective measures to keep the environment pollution-free that is befriendly to all species.

I sincerely thank the Almighty for giving me the mind,

health, time, and life to write this book. I would like to thank my life-cum-wife, Kamala, my son, Jackson Olive, and daughter, Diamond Celestine, for providing me the necessary moral support. Without their active support, this book would not have seen the light of the day. I would also like to thank the publisher who readily expressed acceptance for publishing the book.

Aluri Jacob Solomon Raju

CONTENTS

1

The Greenhouse Effect

The greenhouse effect was discovered by Joseph Fourier in 1824 and was first investigated quantitatively by Svante Arrhenius in 1896. It is the process by which absorption and emission of infrared radiation by atmospheric gases warm a planet's lower atmosphere and surface. The name comes from an incorrect analogy with the warming of air inside a greenhouse compared to the air outside the greenhouse. The Earth receives energy from the Sun in the form of radiation. Most of the energy is in visible wavelengths and in infrared wavelengths that are near the visible range (often called "near infrared"). The Earth reflects about 30% of the incoming solar radiation. The remaining 70% is absorbed, warming the land, atmosphere and oceans. For the Earth's temperature to be in steady state so that the Earth does not rapidly heat or cool, this absorbed solar radiation must be very closely balanced by energy radiated back to space in the infrared wavelengths. The visible solar radiation mostly heats the surface, not the atmosphere, whereas most of the infrared radiation escaping to space is emitted from the upper atmosphere, not the surface. The infrared photons emitted by the surface are mostly absorbed in the atmosphere by greenhouse gases and clouds and do not escape directly to space.

The greenhouse effect can be visualized as follows: Imagine that Earth has been encircled by a giant glass sphere.

The heat of the sun penetrates through the glass. Some of the heat is absorbed by the Earth, and some of it is radiated back towards space. The radiated heat reaches the glass sphere and is prevented from dispersing any further. Similarly, the earth is surrounded by a blanket of gases. This blanket traps energy in the atmosphere, much the same way as glass traps heat inside a greenhouse. This results in an accumulation of energy, and the overall warming of the atmosphere. The 'greenhouse effect' is the popular expression for the above process.

The greenhouse effect is a natural process which made life on Earth possible. Without naturally occurring greenhouse gases such as water vapour, carbon dioxide, methane and nitrous oxide, the Earth's surface temperature would be 33°C cooler - a chilly -18°C rather than the tolerable 15°C. However, since the beginning of industrialization, 200 years ago, concentrations of these gases have increased substantially. It is estimated that the Earth's average temperature has risen by 0.5 to 0.6°C since 1880 because of emissions of greenhouse gases from human activity.

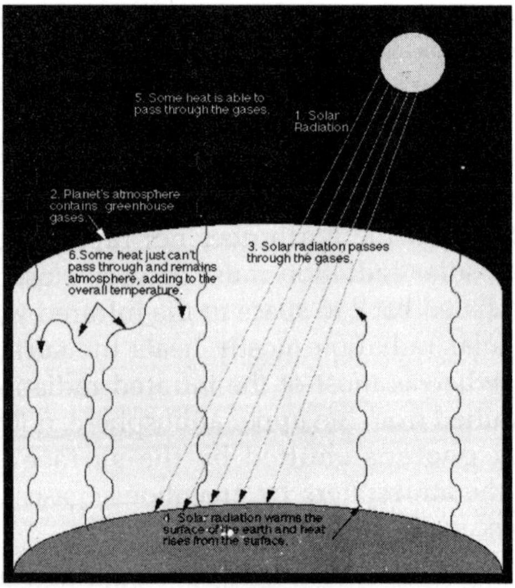

The greenhouse mechanism is an absolute necessity for the survival of the earth. The presence of water vapor in the atmosphere is one of the things that make earth livable for humanity. Carbon dioxide and water vapor trap heat in the atmosphere, a property commonly called the greenhouse effect, while other gases like nitrogen (N_2) and argon (Ar) allow heat to escape to space. The presence of water vapor in the atmosphere helps keep surface air temperatures on the earth range from about -40°C to 55°C. Temperatures on planets without water vapor in the atmosphere, like Mars, stay as low as -100°C.

Once water vapor is in the air, it circulates within the atmosphere. When an air package rises and cools, the water vapor condenses back to liquid water around particulates like dust, called condensation nuclei. Initially, these condensation droplets are much smaller than raindrops and are not heavy enough to fall as precipitation. These tiny water droplets create clouds. As the droplets continue to circulate within the clouds, they collide and form larger droplets, which eventually become heavy enough to fall as rain, snow, or hail. Though the amount of precipitation varies widely over the surface of the earth, evaporation and precipitation are globally balanced. In other words, if evaporation increases, precipitation also increases; rising global temperature is one factor that can cause a worldwide increase in evaporation from the world's oceans, leading to higher overall precipitation.

Since oceans cover around 70% of the earth's surface, most precipitation falls right back into the ocean and the cycle begins again. A portion of precipitation falls on land, however, and it takes one of several paths through the hydrologic cycle. Some water is taken up by soil and plants, some runs off into streams and lakes, some percolates into the groundwater reservoir, some falls on glaciers and accumulates as glacial ice.

The amount of precipitation that soaks into the soil depends on several factors: the amount and intensity of the precipitation, the prior condition of the soil, the slope of the landscape, and the presence of vegetation. These factors can interact in sometimes surprising ways - a very intense rainfall onto very dry soil, typical of the desert south-west, often will not soak into the ground at all, creating flash-flood conditions. Water that does soak in becomes available to plants. Plants take up water through their root systems; the water is then pulled up through all parts of the plant and evaporates from the surface of the leaves, a process called transpiration. Water that soaks into the soil can also continue to percolate down through the soil profile into groundwater reservoirs, called aquifers. Aquifers are often mistakenly visualized as great underground lakes; in reality, groundwater fills the pore spaces within sediments or rocks.

Water that does not soak into the soil collects and moves across the surface as run-off eventually flowing into streams and rivers to get back to the ocean. Precipitation that falls as snow in glacial regions takes a somewhat different journey through the water cycle, accumulating at the head of glaciers and causing them to flow slowly down valleys.

Naturally occurring greenhouse gases have a mean warming effect of about 33 °C, without which Earth would be uninhabitable. On Earth, the major greenhouse gases are water vapor, which causes about 36–70% of the greenhouse effect, carbon dioxide, which causes 9–26%; methane, which causes 4–9%; and ozone, which causes 3–7%. The strength of the greenhouse effect changes with changes in the atmospheric concentrations of some greenhouse gases caused by human activity. Warming by the addition of long-lived greenhouse gases such as carbon dioxide will cause more water to evaporate into the atmosphere. Since water vapor itself acts as a greenhouse gas, the atmosphere warms further; this warming causes more water vapor to evaporate and so on until other processes stop the feedback loop. The result is a much larger greenhouse effect than that due to carbon dioxide alone. Although this process causes an increase in the absolute moisture content of the air, the relative humidity stays nearly constant or even decreases slightly because the air is warmer.

This effect can only be reversed slowly as carbon dioxide has a long average atmospheric lifetime.

Contributors to Greenhouse Effect

Those gas molecules in the Earth's atmosphere with three or more atoms are called "greenhouse gases" because they can capture outgoing infrared energy from the Earth, thereby warming the planet. The greenhouse gases include water vapor with three atoms (H_2O), ozone (O_3), carbon dioxide (CO_2), and methane (CH_4). Also, trace quantities of chlorofluorocarbons (CFC's) can have a disproportionately large effect.

Human activity since the industrial revolution has increased the concentration of various greenhouse gases, leading to increased radiative forcing from carbon dioxide, methane, tropospheric ozone, chlorofluorocarbons and nitrous oxide. Molecule for molecule, methane is a more effective greenhouse gas than carbon dioxide, but its concentration is much smaller so that its total radiative forcing is only about a fourth of that from carbon dioxide. Some other naturally occurring gases contribute very small fractions of the greenhouse effect; one of these, nitrous oxide, is increasing in concentration owing to human activity such as agriculture. The atmospheric concentrations of carbon dioxide and methane have increased by 31% and 149% respectively since the beginning of the industrial revolution in the mid-1700s. These levels are considerably higher than at any time during the last 6, 50,000 years. The present atmospheric concentration of carbon dioxide is about 385 parts per million by volume. Future carbon dioxide levels are expected to rise due to ongoing burning of fossil fuels and land-use change. The rate of rise will depend on uncertain economic, sociological, technological, and natural developments, but may be ultimately limited by the availability of fossil fuels. It is expected that the future emissions may range from 541 to 970 ppm by the year 2100. Fossil fuel reserves are sufficient to reach this level and continue emissions past 2100, if coal, tar sands or methane clathrates are extensively used.

The principal greenhouse gases that enter the atmosphere because of human activities are the following:

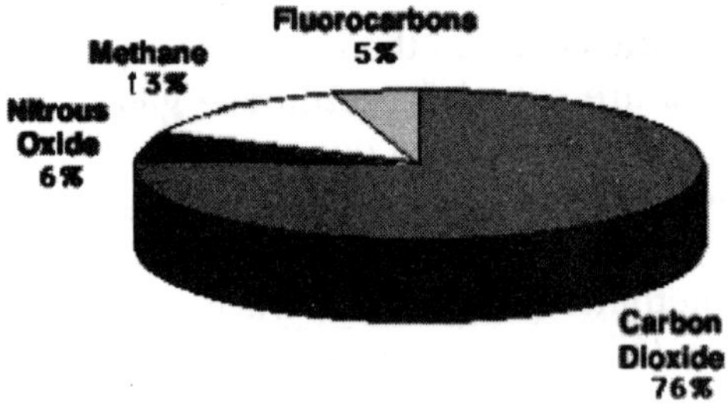

Distribution of Greenhouse gases in Earth's atmosphere

Carbon dioxide

It is a colorless, odorless non-flammable gas and is the most prominent greenhouse gas in Earth's atmosphere. It is recycled through the atmosphere by the process of photosynthesis, which makes human life possible. Photosynthesis is the process of green plants and other organisms transforming light energy into chemical energy. Light Energy is trapped and used to convert carbon dioxide, water, and other minerals into oxygen and energy rich organic compounds. Carbon dioxide is emitted into the air as humans exhale, burn fossil fuels for energy, and deforest the planet. Every year humans add over 30 billion tons of carbon dioxide in the atmosphere by these processes, and it is up thirty per cent since 1750. Fossil Fuels were created chiefly by the decay of plants from millions of years ago. Coal, oil and natural gas are used to generate electricity, heat our homes, power our factories and run our cars. These fossil fuels contain carbon, and when they are burned, they combine with oxygen, forming carbon dioxide. The two atoms of oxygen add to the total weight. Global carbon dioxide emissions from burning fossil fuels rose 12% between 1990 and 1995. The increase from developing countries was three times that from developed countries. Middle East carbon dioxide emissions from burning of fossil fuels increased 35%, Africa increased 12%, and Eastern Europe increased rates by 75% from 1990-1995.

In 2003, the US emitted 5.8 billion metric tons of carbon dioxide. Western Europe - the EU 15 plus Switzerland, Iceland, Norway, and Turkey - had emissions of 3.9 billion metric tons. Japan emitted 1.2 billion metric tons, China emitted 3.5 billion metric tons, and India emitted just over 1 billion. On a per capita basis, the disparity is apparent: In 2003, the US emitted 20 tons per person per year, compared to 2.7 tons per person in China, and 0.96 tons per person in India.

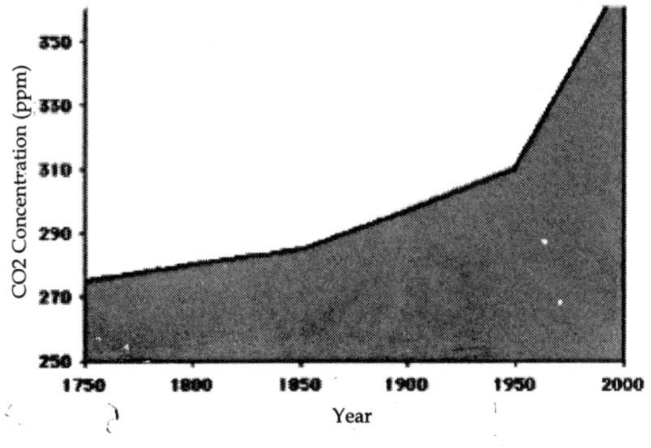

The increase of carbon dioxide in the air over the past few centuries

Deforestation is another main producer of carbon dioxide. The causes of deforestation are logging for lumber, pulpwood and fuel wood. Also contributing to deforestation is clearing new land for farming and pastures used for animals such as cows. Forests and wooded areas are natural carbon sinks. This means that as trees absorb carbon dioxide, and release oxygen, carbon is being put into trees. This process occurs naturally by photosynthesis, which occurs less and less as trees are cut and burnt down. As the abundance of trees declines, less carbon dioxide can be recycled. When trees are burnt, carbon is released into the air and the carbon bonds with oxygen to form carbon dioxide, adding to the greenhouse effect. Carbon dioxide enters the atmosphere through the burning of oil, natural gas, coal, solid waste, trees and wood products, and also as a result of other chemical reactions. Carbon dioxide is also removed from the atmosphere when it is absorbed by plants as part of the biological carbon cycle.

Methane

It is a colorless, odorless, flammable gas. It is formed when plants decay and where there is very little air. It is often called swamp gas because it is abundant around water and swamps. Bacteria that breakdown organic matter in wetlands and bacteria that are found in cows, sheep, goats, buffaloes, termites, and camels produce methane naturally. Since 1750, methane has doubled, and could double again by 2050. Each year 350-500 million tons of methane is added to the air by raising livestock, coal mining, drilling for oil and natural gas, rice cultivation, and garbage sitting in landfills. It stays in the atmosphere for only 10 years, but traps 20 times more heat than carbon dioxide.

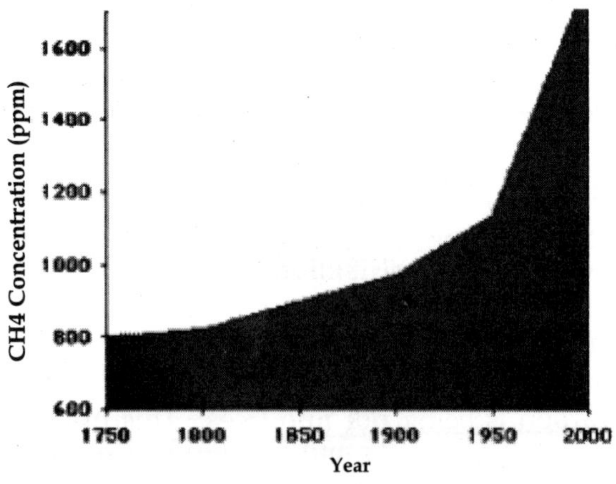

Methane is on the rise since 1750

Rice cultivation has developed into a large business; farmland has doubled in the past 45 years. It feeds 1/3 of the

World's population. It grows mostly in flooded fields, where bacteria in water-logged soil release methane. Livestock such as cows, sheep, goats, camels, buffaloes, and termites release methane as well. Bacteria in the gut of the animal break down food and convert some of it to methane. When these animals belch, methane is released. In one day, a cow can emit ½ pound of methane into the air. Methane is emitted during the production and transport of coal, natural gas, and oil. Methane emissions also result from livestock and other agricultural practices and by the decay of organic waste in municipal solid waste landfills.

Nitrous oxide

It is another colorless greenhouse gas; however, it has a sweet odor. It is primarily used as an anesthetic because it deadens pain and for this characteristic is called laughing gas.

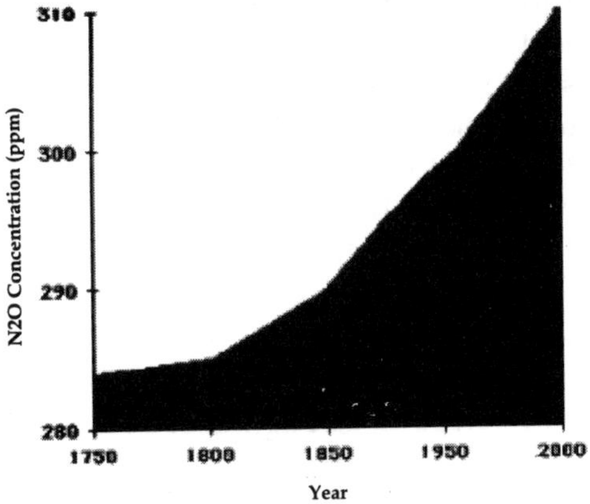

Nitrous Oxide has been on the rise since 1750

This gas is released naturally from oceans and by bacteria in soils. Nitrous oxide gas rose by more than 15% since 1750. Each year we add 7-13 million tons into the atmosphere by using nitrogen based fertilizers, disposing of human and animal waste in sewage treatment plants, automobile exhaust, and other sources not yet identified. It is important to reduce emissions because the nitrous oxide released today will still be trapped in the atmosphere 100 years from now.

Nitrogen based fertilizer use has doubled in the past 15 years. These fertilizers provide nutrients for crops; however, when they breakdown in the soil, nitrous oxide is released into the atmosphere. In automobiles, nitrous oxide is released at a much lower rate than carbon dioxide, because there is more carbon in gasoline than nitrogen. Nitrous oxide is emitted during agricultural and industrial activities, as well as during combustion of fossil fuels and solid waste.

Flourinated gases

It is a general term for any group of synthetic organic compounds that contain fluorine and carbon. Hydrofluorocarbons, perfluorocarbons, and sulfur hexafluoride are synthetic, powerful greenhouse gases that are emitted from a variety of industrial processes. Fluorinated gases are sometimes used as substitutes for ozone depleting substances such as CFCs, HCFCs and halons. Many of these compounds, such as chlorofluorocarbons, can be easily converted from gas to liquid or liquid to gas. Because of these properties, chlorofluorocarbons can be used in aerosol cans, refrigerators, and air conditioners. When chlorofluorocarbons are emitted into the atmosphere, they break down molecules in the Earth's ozone layer. These gases are typically emitted in smaller quantities, but because they are potent greenhouse gases, they are sometimes referred to as High Global Warming Potential gases.

Since the industrial revolution about 850 billion tons of CO_2 have been emitted due to combustion of fossil fuels, oil, coal and natural gas. An additional 370 billion tons have been added through changes in land use and deforestation. Every year humans emit around 25 billion tons of CO_2 into the atmosphere, which equals approximately 48,000 tons every minute. Some estimates show that a reservoir of 37, 000 billion tons of CO_2 is buried in the ground as oil, coal and gas. The largest portion of this CO_2 reservoir is in coal, and the second largest is in natural gas. This is a chilling reminder of what is in store for the planet, should we decide to continue to burn fossil fuels without any restrictions. Experts estimate that emission of 2,500 billion tons of CO_2 - a mere 7% of the existing stock of fossil fuels - will result in a doubling of pre-industrial concentrations of CO_2.

Greenhouse gas emissions represent waste. They are the result or the byproduct of something. Carbon dioxide is the by product of people breathing. Carbon monoxide is the by product of combustion engines like the one that runs your car. Methane is produced naturally by rotting vegetation and, believe it or not, by the solid waste of all kinds of animals. These make up just a few of the greenhouse gases. While some are not a big deal when they occur in moderate amounts, the massive accumulation of them in the atmosphere is what is causing the problem.

The green house effect only started being a problem when an excess of greenhouse gases was generated. In the last one hundred years, this has caused two important phenomena: the average global temperature has risen by 0.3 to 0.6°C, while in the same period the sea level has risen by 10 to 25 cm. The second phenomenon automatically follows from the first one: due to heating, the water in the oceans expands, thus taking up more space, and furthermore the ice of glaciers and the polar ice caps melts. For example, if it gets too warm on our planet, glaciers are about to thaw and vanish. Their fauna is also in danger, like white bears or some fishes, which cannot adapt to warmer water. These glaciers, if they thaw, will make the level of oceans increase, and it could create more and more inundations.

Concerning the climate, periods of dryness and scorching heat will be more frequent, which is unhealthy, especially for old people and babies, who are weaker. These temperatures could also have bad effects on agriculture and crops. In poor countries, where, unfortunately, food is already rare, it will be a serious problem for people's survival. Moreover, meteorological accidents like hurricanes, twisters... or storms could be part of our life, and we already know their terrible effects. It always causes material damages and human losses.

The man-made greenhouse effect is currently in the process of changing those conditions in the long term, without our being able to control the consequences, one of which is likely to be the continued steep increase in extreme weather events. Conditions locally may differ from global averages in temperature, amounts of rain and snow, frequency or severity of major storms. Local conditions like these profoundly influence the quality of our lives, and they certainly affect important activities like agriculture. Farmers need to know, for example, whether the amounts of soil moisture available for their crops will increase or decrease and whether they should worry more or less about frosts. In some regions, concerns for safety and property could require measures to deal with more frequent flooding, more severe storms, or the like. Nevertheless, they cannot yet produce good estimates of future changes in regional climate because their tools are not yet sophisticated enough to forecast changes in regional conditions.

With this point of view, it is necessary to make something to change the situation. The matter is that it will not change in one day, thus the present generations must not be selfish. The future generations will suffer more than the present generations do, but the present generations have to make their future life easier. That is why the present generations are obliged to act today and all the days following. Solutions must be found and applied nowadays, in order to limit bad effects for the future.

Green Revolution and Greenhouse Gases

The term "Green Revolution" was first used in 1968 by former USAID director William Gaud, who noted the spread of the new technologies and said, "These and other developments in the field of agriculture contain the makings of a new revolution. It is not a violent Red Revolution like that of the Soviets, nor is it a White Revolution like that of the Shah of Iran. It is Green Revolution. There were three basic elements in the method of the Green Revolution - continuing expansion of farming areas, double-cropping in the existing farmland and using seeds with improved genetics. Introduction of improved seed strains, fertilizers, and irrigation as a means of producing higher yields in crops such as rice, wheat, and corn; particularly important in the densely populated countries of Asia.

The world's worst recorded food disaster occurred in 1943 in British-ruled India. Known as the Bengal Famine, an estimated 4 million people died of hunger that year in eastern India (which included today's Bangladesh). Initially, this catastrophe was attributed to an acute shortfall in food production in the area. When the British left India in 1947, India continued to be haunted by memories of the Bengal Famine. It was therefore natural that food security was one of the main items on free India's agenda. This awareness led, on one hand, to the Green Revolution in India and, on the other, legislative measures to ensure that businessmen would never again be able to hoard food for reasons of profit.

The Green Revolution, spreading over the period from 1967/68 to 1977/78, changed India's status from a food-deficient country to one of the world's leading agricultural nations. Until 1967 the government largely concentrated on expanding the farming areas. But the population was growing at a much faster rate than food production. This called for an immediate and drastic action to increase yield. The action came in the form of the Green Revolution. The term 'Green Revolution' is a general one that is applied to successful agricultural experiments in many developing countries. India is one of the countries where it was most successful.

The area of land under cultivation was being increased from 1947 itself. But this was not enough to meet the rising demand. Though other methods were required, the expansion of cultivable land also had to continue. So, the Green Revolution continued with this quantitative expansion of farmlands.

Double cropping was a primary feature of the Green Revolution. Instead of one crop season per year, the decision was made to have two crop seasons per year. The one-season-per-year practice was based on the fact that there is only one rainy season annually. Water for the second phase now came from huge irrigation projects. Dams were built and other simple irrigation techniques were also adopted.

Using seeds with superior genetics was the scientific aspect of the Green Revolution. The Indian Council for Agricultural Research (which was established by the British in 1929) was reorganized in 1965 and then again in 1973. It developed new strains of high yield variety seeds, mainly wheat and rice and also millet and corn.

The Green Revolution was a technology package comprising material components of improved high yielding varieties of two staple cereals (rice and wheat), irrigation or controlled water supply and improved moisture utilization, fertilizers, and pesticides, and associated management skills.

The Green Revolution resulted in a record grain output of 131 million tons in 1978/79. This established India as one of the world's biggest agricultural producers. Yield per unit of farmland improved by more than 30% between 1947 (when India gained political independence) and 1979. The crop area under high yielding varieties of wheat and rice grew considerably during the Green Revolution. The Green Revolution also created plenty of jobs not only for agricultural workers but also industrial workers by the creation of related facilities such as factories and hydroelectric power stations.

In spite of this, India's agricultural output sometimes falls short of demand even today. India has failed to extend the concept of high yield value seeds to all crops or all regions. In terms of crops, it remains largely confined to food grains only, not to all kinds of agricultural produce. The Green Revolution has created some problems mainly to adverse impacts on the environment. The increasing use of agrochemical-based pest and weed control in some crops has affected the surrounding environment as well as human health. Increase in the area under irrigation has led to rise in the salinity of the land. Although high yielding varieties had their plus points, it has led to significant genetic erosion.

These 'miracle' seeds were designed to overcome the limits placed on chemically intensive agriculture by the indigenous seeds. They became central to breaking out of nature's limits and cycles. The 'miracle seeds' of the green revolution transformed the common genetic heritage into private property, protected by patents and intellectual property rights. The green revolution changed the 10,000-year evolutionary history of crops by changing the fundamental nature of seeds. It involved the use of chemical fertilizers, pesticides, mechanization and petroleum, intensive and accurate irrigation which is made possible by building of dams. High yields are not intrinsic to the seeds, but are a function of the availability of required inputs, which in return have ecologically destructive impacts. Land degradation is caused by water logging, salinization of the soil, desertification and water scarcity, destroying water resources, destruction of soil fertility, micronutrient deficiency, soil toxicity by high use of pesticides and chemical fertilizers and biomass reduction used for fodder and organic manure.

The green revolution functions properly only when the physical environment, especially the availability of water is sufficient. This higher need for water is caused by the shift from water prudent crops such as millets and oilseeds to monocultures and multicropping such as wheat and rice and by the replacement of old varieties of wheat and rice with new varieties of wheat and rice. Intensive irrigation is required mainly by building large dams and applying surface irrigation. The dramatic increase in water use has led to a total destabilization of the water balance. The water cycle can be destabilized by adding more water to an ecosystem than the natural drainage potential of that system. This leads to desertification through water logging and salinization of the land. Desertification of this kind is a form of water abuse rather than water use. The problem salinity arises through intensive irrigation in arid regions. In regions of scarce rainfall, the earth contains a large amount of unleached salts. Pouring irrigation water into such soils brings those salts to the surface and leaves behind a residue when the water evaporates. Where irrigation is dependent on ground water, the water table is declining at an estimated rate of one to one and a half foot every year, due to over-exploitation. The nutrient cycle, in which nutrients are produced by the soil through plants, and returned to the soil as organic matter is thus replaced by linear non-renewable flows of phosphorous and potash derived from geological deposits, and nitrogen derived from petroleum. This led to a western NPK-mentality, with high gifts of these three minerals. As a result deficiency of micronutrients as zinc, iron, copper, manganese and magnesium arose. Soil toxicity arose through irrigation and high chemical fertilizer input, for example fluorine, boron, selenium and aluminium toxicity. It is posing a threat to crop production as well as animal health'. As a result of the reductionist approach only the output of crops were counted, but not the loss in maintaining the conditions of productivity. These outputs have also to be uniform like the central market wants them, and not divers like the traditional crops which partly were used for own food. The indigenous cropping systems are based on internal organic inputs. They used straw from the harvested crops and other through westerns considered wastes to feed the farm animals, and/or to increase soil fertility. Also the animals provided organic manure. The new varieties were however selected by producing little straw, because otherwise as a result of high fertilizer input they would lodge and the crop would be lost. So the straw production is much lower, with negative effects on soil fertility through lower input of biomass. The heavy use of chemical fertilizers and new seeds directly led to decreased soil fertility, because the soil productivity (which also needs organic mass) is lowered and the nutrient recycling is disturbed. The results are new deficiencies and diseases. As a result of replacing pulses by wheat and rice the nitrogen fixing capacity of crops was lost, so more fertilizers are needed. This process implies a downward spiraling of agricultural land use from legume to wheat to rice to wasteland. Genetic erosion caused by mixtures and rotation of diverse crops like wheat, maize, millets, pulses and oil seeds replaced by monocultures of wheat and rice; the introduced wheat and rice varieties reproduced over large-scale as monocultures came from a very narrow genetic base, compared to the high genetic variability in the populations of traditional wheat or rice plants which resulted in explosive growth of pests in the crops.

In addition to these negative effects, green revolution contributed to the release of greenhouse gases such as carbon dioxide, methane and oxides of nitrogen. The excessive use of chemical fertilizers, NO_2 is released, which is a greenhouse gas, as is CO_2 produced by more use of mechanization instead of hand labour. World cattle population has doubled and accordingly the release of methane from their rumen has proportionately occurred. Bacteria that break down cellulose in the guts of cattle convert between 3 and 10 per cent of the food that the cattle eat, into methane. Wherever bacteria break down organic matter in the absence of oxygen, they produce methane. When the same process occurs in the presence of oxygen, carbon dioxide is produced. Concentrations of methane in the air have been rising at 1 per cent per year, at least since 1950. This is four times the rate of increase of carbon dioxide. Levels are already more than double those recorded before the explosion in human activity on Earth that followed the Industrial Revolution. There are 250, 000 billion termites in the world, occupying two-thirds of the world's land area and eating a third of the world's vegetation. The annual emission of methane from termites could amount to 150 million tons, almost a third of the total. And the figure could be rising fast. Termites like grasslands best, so as farmers replace tropical rainforests with pasture, the termite population could be growing. Production could already be around 70 million tons a year and very large increases in methane production from waste dumps are expected in the coming decades from the developing world. Paddies could be extremely efficient creators of atmospheric methane. At certain seasons and times of day, the roots of rice plants seem to capture methane from the muddy bottoms and transport it through the plant's vascular system and into the air, thus bypassing microorganisms in the water that would reoxidise some of the methane. Up to 90 per cent of methane from the depths of the flooded fields seems to reach the air this way. The green revolution is fundamentally changing the rice paddies of the world, notably by the use of fertilizers. The urea and ammonium sulphate reduce methane emissions, while other fertilizers, including rice straw, double it. India, with its millions of cattle and swathes of paddy fields, is sometimes called the biggest source of methane in the world. Much the same is true for the output of methane from burning trees and grasslands. Methane currently lasts an average of 10 years in the atmosphere. After that, it may be consumed by oxidizing bacteria or by chemical processes in the atmosphere itself. Bacteria that oxidize methane turn up in marine sediments, lakes and the water table in wetlands, but probably the biggest sink on land is bacteria in soils. Another big unknown is the oceans. There appears to be little exchange of methane between the air and the sea. But that is because methane rising from the ocean depths is apparently oxidized by marine organisms and so never reaches the air.

These greenhouses gases polluted the atmosphere. This revolution puts new demands on non-renewable resources like fuel and new demands on scare renewable resources like water. Further, it resulted in poverty for the local population in the form of indebted farmers, landless farmers, unemployed or only seasonally employed land labourers and building dams or irrigation canals.

Lifestyles and Emissions of Greenhouse Gases

Lifestyle is defined as "products and services consumed as part of an individual's day to day life and over which he or she has a degree of choice regarding product, quantity and price ". Each person's lifestyle is therefore developed by themselves in daily negotiation with the wider society and the consumer market. An individual's lifestyle consumption is composed of house-heating, lighting, appliances; transport-cars, buses, flights; and production consumption-clothes, food, and general household "goods". Manufacturing a car produces 5 tons of carbon dioxide. After that, every liter of petrol produces 2.5 kg of carbon dioxide, on average consumption that is just less than half a kilo of carbon dioxide per mile. Someone who commutes 20 miles each way to work will produce over 4 tons of carbon dioxide from commuting alone. Flights are ignored in national and international climate change statistics and are therefore ignored in national accounting of greenhouse gas emissions. Jet planes drink fuel, so even though they are a form of public transport, the oil consumption per person per mile is approximately the same as a car with one person in it. However, the greenhouse impact is three times as high per mile because jet planes distribute pollution and water vapour high in the atmosphere. One long haul flight can outweigh all other personal emissions. Typical return flights produce carbon dioxide as follows: London-Paris return flight produces 626 kg carbon dioxide; London-Sydney flight produces 10.8 tons carbon dioxide; London-New York flight produces 4.25 tons carbon dioxide.

Some greenhouse gases are part of vital and basic lifestyles – meaning they are necessities, not luxuries – and not much can be done about them. For example, keeping cattle and sheep for milk, meat or wool causes greater quantities of methane in the atmosphere, as does paddy rice farming. But this is how most of Asia lives, and stopping rice cultivation and cattle rearing would mean having to find new sources of food and income.

The biggest greenhouse gas culprit, brought on by human activity, is carbon dioxide. Carbon dioxide comes from a wide variety of sources, but mostly from burning organic matter. Anything that is powered by fuel such as petrol, diesel, coal or even wood, releases carbon dioxide. The more we depend on our cars, planes and trains, the more we release carbon dioxide into the atmosphere. Our industries are powered by these fuels, as is agriculture that uses petroleum to power machinery, produce pesticides and fertilizer, and transport produce to market.

Electricity-generating plants powered by coal or oil, as are many in India, also release carbon dioxide. The world is now what is called a 'carbon economy', meaning that our worldwide economic system is based on burning petroleum or coal – known as 'fossil fuels' – and there seems to be no way of getting around this. Our transport, industry, food, heating and cooling, household electricity that powers all our appliances – all are based on deriving power from these sources.

Plants use atmospheric CO_2 to build their bodies, the way we use food. This means that trees can store large amounts of carbon dioxide as they grow. Unfortunately, however, we are not planting enough trees, and are cutting down whatever trees there already are. The more forests we destroy, the more carbon dioxide we release into the atmosphere, compounding the problem caused by our use of fossil fuels. And it is the most industrialized and developed countries that create the most greenhouse gases, because they are the ones that burn the most fossil fuels, with their heavy use of cars and other energy-intensive devices. These are 'luxury' emissions of carbon dioxide, not essential ones, since people can do without their cars and planes. Developed countries also cut down the most forests (in their own country or in others) to fuel their industrial growth.

General consumption can be very significant. It encompasses all agricultural emissions, most industry and most transport. The general consumption of average American produces 10 tons of carbon dioxide. The UK average is probably around half this is around 4-5 tons per year. Diet is also a factor- home produced vegetables are carbon neutral, average supermarket vegetables have often used up their own weight in fossil fuels. To produce one pound of beef, a cow has produced half a pound of methane, a very powerful greenhouse gas which is equivalent to 10.5 lbs of carbon dioxide. The beef eaten by the average American in a year has produced the methane equivalent of 1.4 tons of carbon dioxide. The average UK person's emissions are ten times higher than an Indian who is effectively subsidizing our emissions. There are also enormous differences between people's emissions within the UK depending on income, size of house, use and size of car. Again the low consumers are subsidizing the impacts of the high consumers. However, poor people can also be high consumers of energy.

Food has always been at the top of the sustainable development agenda focusing on prominent issues such as food security, land rights, soil degradation and trading relationships. Now there is a growing recognition that how we grow, distribute, store, sell and consume food also makes a significant contribution to one of the greatest sustainability threats of all - climate changing greenhouse gas emissions.

Globally, the main greenhouse gas (GHG), carbon dioxide, accounts for about 72 per cent of all GHG emissions. However there are other GHGs, and these are particularly associated with the food chain. Methane, much of which comes from grazing livestock, contributes 20 per cent while arable soils and livestock wastes contributes to the amount of nitrous oxide which, in turn, adds another seven per cent to the total global warming effect. Greenhouse gases emit at all stages in the food chain including agricultural production, manufacturing, refrigeration, transport, packaging, retail, home storage, cooking and waste disposal. A change in practice at one point in the food chain can positively or negatively affect emissions at another stage.

Agriculture accounts for almost half of food-related greenhouse gas emissions. It is agricultural production, rather than food transport, that poses the major challenge. Grazing livestock together with rice cultivation are responsible for major fluxes of methane, while both arable soils and livestock wastes are potent sources of nitrous oxide. While the agricultural stage of food production represents the main cause for concern, the importance of other supply chain stages may be greater for some particular foods. Potatoes, chickpeas and tea leaves, for example, are less greenhouse gas intensive to grow they are to cook. Think of cooking a baked potato in an oven, boiling chickpeas for an hour till soft, or switching the kettle on for tea. For frozen vegetables, the refrigeration stage will dominate as an 'emission hotspot.' In the developed world refrigeration is used at almost every stage in the supply chain and plays a vital role in reducing waste. This is up to the point where people bring food home; after that, cultural attitudes, cheap food and attitudes to food wastage mean up to a third of household food, most of it edible, is thrown away. This is a major concern since wasted food represents a waste of all the embedded GHG emissions associated with growing, distributing and selling that food. Where it can prevent food waste, refrigeration helps avoid the emission of greenhouse gases. In the developing world, post harvest losses are greater because there is less access to refrigeration and clearly more and better refrigeration could help reduce spoilage. But refrigeration comes with its own greenhouse gas burdens because of the refrigerant gases used which has a high global warming potential and, more significantly, due to the energy they require. Some refrigeration use is unavoidable and desirable. However its ready availability means that a 'refrigeration dependence' culture has developed. The types of foods we consume are now predicated on refrigeration's existence. A warming global climate might also increase our reliance on, and need for, refrigeration particularly if we are not prepared to accept changes to the sorts of foods we consume.

All this very clearly goes against current and emerging patterns of consumption. In the developed world, our food system is predicated on the availability of cheap food, and particularly cheap meat and dairy products, provided by a sophisticated and economically efficient manufacturing and retail infrastructure. The signs are that the rapidly growing middle classes in the developing world aspire to similar lifestyles and a food industry is rapidly developing to cater to such aspirations. At present, there are few indications that policy makers anywhere in the world are willing to tackle excessive consumption although some are at least starting to talk cautiously about the issue. The danger is that if we do not take steps now to address what and how much we consume, a changing climate may force altered patterns of consumption on us. And if this arises, the fear is that those who have very little already will, as ever, stand to lose the most.

The food system has an impact on climate change but the reverse is also true. A warming climate is likely to present major challenges to current systems of agricultural production and food provisioning. Poorer parts of the world that tend to be in the lower latitudes are likely to experience the worst effects of climate change, including drought or flooding, or both. Richer countries in Northern Europe and North America may find that initially warmer temperatures and longer growing seasons actually favour agricultural production, but by 2050 the negative impacts of high temperatures and water shortages will start to hit. Climate change could also trigger changing patterns of crop and livestock disease while extreme weather events could disrupt transport systems and stationary infrastructure.

Industrialized activities such as fossil fuel burning, cement production and tropical deforestation have increased the carbon dioxide concentrations in the atmosphere, and industrialized agricultural practices such as factory farming have led to a significant increase of methane from overproduction of living organisms who release this gas, such as sheep and cattle. More greenhouse gases in the atmosphere basically means more Infrared absorption and recycling of that energy back down to Earth, where it raises the Earth's temperature. While industry in general is often blamed directly for these emissions, sometimes the consumer forgets that he or she is indirectly responsible as well, by purchasing their products and refusing to relinquish or scale back on current lifestyles that contribute to the problem.

Deforestation

The world currently has about ten billion acres of forest. Deforestation is caused by exploitation of natural resources—including expanding populations, logging, agriculture, biofuel production, and wildfires. The world's forests store 283 gigatons of carbon in their biomass alone, while the total carbon stored in forest biomass, deadwood, litter and soil together is roughly fifty per cent more than the amount found in the atmosphere – adding up to one trillion tons. The destruction of forests adds almost two billion tons of carbon to the atmosphere each year. Preventing this stored carbon from escaping is important for maintaining the global carbon balance and vital to conserving the environment. Particularly in the tropics, where vegetation grows rapidly and therefore pulls carbon from the atmosphere more quickly, planting trees can remove large amounts of carbon dioxide from the air within a relatively short time. There, forests can fix as much as 15 tons of carbon in their biomass and wood per hectare per year.

Trees generally are about 20 per cent carbon by green weight, and in addition to the trees themselves, the overall biomass of forests also acts as a "carbon sink." For instance, the organic matter in forest soils – like the humus produced by the decomposition of dead plant material – also acts as a very large carbon store that can store carbon for thousands of years. As a result, forests store enormous amounts of carbon: all in all, the world's forests and forest soils currently store more than one trillion tons of carbon – about one and a half time the amount found floating free in the atmosphere.

Carbon emissions from tropical deforestation have long been recognized as a key component of the global carbon budget, and more recently of our global climate system. Tropical forest clearing accounts for roughly 20% of anthropogenic carbon emissions and destroys globally significant carbon sinks. Loss of forests contributes as much as 30 per cent of global greenhouse gas emissions each year—rivaling emissions from the global transportation sector. Almost 20 per cent of global greenhouse gas emissions comes from clearing the world's forests. This is more than all of the world's emissions from transport - second only to emissions produced from burning fossil fuels to produce electricity. Therefore, deforestation in all parts of the world is a major concern for the release of greenhouse gases, particularly carbon. Further, deforestation opens the soil to direct sunlight and the soil organic matter gets oxidized and releases carbon into the atmosphere. In effect, the carbon stored in forest trees and soil is simply kept in the atmosphere by unwise human activities in the form of deforestation.

In India, deforestation seems to be a routine daily activity in almost all parts of the country. Reforestation in the form of different schemes is also being carried out by the forest departments in order to maintain sizeable forest cover in order to sustain humanity in this part of the world. But, reforestation with only a small bunch of species is an unfortunate thing to recover the forest cover loss. Further, deforestation is expanding rapidly due to promotion of urbanization, industrialization and better road and rail transportation facilities to most part of the country. Agricultural lands are also being converted for setting up industries, residential complexes, universities, corporate colleges, information technology parks, etc. In this process, there is gradual decline in the area of forest cover and added to this, there is loss of species diversity. Deforestation in these different forms is promoting the release of greenhouse gases, especially carbon into the atmosphere.

Hydroelectric Dams

Hydroelectric dams produce significant amounts of carbon dioxide and methane, and in some cases produce more of these greenhouse gases than power plants running on fossil fuels. Carbon emissions vary from dam to dam. This is because large amounts of carbon tied up in trees and other plants are released when the reservoir is initially flooded and the plants rot. Then after this first pulse of decay, plant matter settling on the reservoir's bottom decomposes without oxygen, resulting in a build-up of dissolved methane. This is released into the atmosphere when water passes through the dam's turbines. The dams produce more methane than the river because of the lack of oxygen.

Seasonal changes in water depth mean there is a continuous supply of decaying material. In the dry season, plants colonize the banks of the reservoir only to be engulfed when the water level rises. For shallow-shelving reservoirs these "drawdown" regions can account for several thousand square kilometers. In effect man-made reservoirs convert carbon dioxide in the atmosphere into methane. This is significant because methane's effect on global warming is 21 times stronger than carbon dioxide's.

Methane from India's dams could boost the country's official greenhouse emissions by 40%. Carbon-containing vegetation in Indian dams is causing an annual release of methane equivalent to 825 million tons of carbon dioxide.

India currently contributes around 4% of global emissions due to rising use of fossil fuels. India has the largest number of dams in the world after the US and China and has constructed many to service its farm sector, which makes up around 22% of the country's Gross Domestic Product and employs 70% of the workforce.

The dams have also been used to help power industry and bring electricity to some of the country's 1.1 billion people, most of who live in villages. But the dam constructions have frequently sparked protests for displacing tens of thousands of poor people, ravaging prime forests housing rare flora and fauna as well as destroying river ecosystems. Gases released from rotting trees that were flooded when dams are formed account for only a fraction of the emissions from reservoirs. Carbon dioxide and methane are released as organic matter continues to wash into the reservoir from upstream and break down. This means that the emissions do not disappear when the flooded forest has broken down but may continue for the lifetime of the reservoir.

In India, the problem of siltation of reservoirs is severe

because of deforestation in upstream areas. Further, the rivers carry huge organic load and it settles at the bottom of reservoirs and gradually break down in the absence of oxygen due to the action of methanogenic bacteria. In this way, methane and carbon dioxide gases are continually released into the atmosphere.

Coal Power Plants

Coal is an important source for electricity production from thermal power plants. Coal burning in power plants releases carbon dioxide along with other pollutants. But, carbon dioxide is the major concern as it happened to be a greenhouse gas. In the world, coal-fired power plants are many and so is the emission of more carbon dioxide. The rate of carbon dioxide release into the atmosphere depends on the quality and composition of coal. There are different forms of coal with different quantities of impurities but major content is carbon. Coal mining also contributes to the release of carbon dioxide and methane gas into the atmosphere. Therefore, coal burning for power production is a serious concern to handle the levels of greenhouse gases which cause global warming effect.

The total installed capacity of the power sector in India is around 1, 28,000 MW. Thermal power plants—using coal, gas or oil—together account for more than 65.6 per cent of the total power capacity in the country. Historically, the Indian power sector has been dominated by coal as the predominant fuel source of power, accounting for 82.45 per cent of the installed capacity followed by gas (16.12 per cent) and oil (1.43 per cent). India has emerged as the fifth largest power market in the world. It ranked eighth in the last decade.

Among thermal-based power generation sources, coal-based power plants rank highest in air pollution, waste generation and water consumption. It is also the largest emitter of carbon dioxide, a greenhouse gas. A typical 500 MW thermal power plant using coal emits around 105 tons per day of SO_2 (at 100 per cent load factor, 0.7 per cent sulphur content in coal), 24 tons per day of NO_2 and 2.5 tons per day of particulate matter (at 34 per cent ash content, 99.9 per cent electrostatic precipitator efficiency) and ash around 3,000-3,500 tons per day. Moreover, disposing one ton of fly ash requires around 1 sq m of land. Indian thermal power plants consume on an average 80 cubic meter of water for every MW-hour (MWh), while the global best practice is just 10 cubic m/MWh.

Carbon dioxide emissions from electric power generation are influenced by the efficiency with which fossil fuels are converted into electricity. In a typical power plant, about one-third of the energy contained in the fuel is converted into electricity, while the remainder is emitted as waste heat. Substantial improvements in generation efficiency can be achieved in the future through the replacement of traditional power generators with more efficient technologies, such as combined-cycle generators and combined heat and power systems. In these types of systems, waste heat is captured to produce additional kilowatt-hours of electricity or displace energy used for heating or cooling. Both strategies result in lower carbon dioxide emissions.

Fuel Wood Combustion

Biomass burning is a source of greenhouse gases, carbon dioxide, methane, and nitrous oxide. In addition, biomass burning is a source of chemically active gases, including carbon monoxide, non-methane hydrocarbons, and nitric oxide. These gases, along with methane, lead to the chemical production of tropospheric ozone (another greenhouse gas) as well as control the concentration of the hydroxyl radical, which regulates the lifetime of almost every atmospheric gas. Following biomass burning, biogenic emissions of nitrous oxide, nitric oxide, and methane are significantly enhanced. It is hypothesized that enhanced post-burn biogenic emissions of these gases are related to fire-induced changes in soil chemistry and/or microbial ecology. Biomass burning, once believed to be a tropical phenomenon, has been demonstrated by satellite imagery to also be a regular feature of the world's boreal forests.

Biomass burning includes the combustion of living and dead material in forests, savannas, and agricultural wastes, and the burning of fuel wood. Under the ideal conditions of complete combustion, the burning of biomass material produces carbon dioxide (CO_2) and water vapor (H_2O), according to the reaction

$$CH_2O + O_2 - CO_2 + H_2O$$

where CH_2O represents the average composition of biomass material. Since complete combustion is not achieved under any conditions of biomass burning, other carbon species, including carbon monoxide (CO), methane (CH_4) non-methane hydrocarbons, and particulate carbon, result through the incomplete combustion of biomass material. In addition, nitrogen and sulfur species are produced from the combustion of nitrogen and sulfur in the biomass material.

While CO_2 is the overwhelming carbon species produced by biomass burning, its emissions into the atmosphere resulting from the burning of savannas and agricultural wastes are largely balanced by its reincorporation back into biomass via photosynthetic activity within weeks to years after burning. However, CO_2 emissions resulting from the burning of forests converted to non-forested areas and other carbon combustion products from all biomass sources including CH_4, CO, Non-methane hydrocarbons, and particulate carbon, are largely "net" fluxes into the atmosphere since these products are not reincorporated into the biosphere.

The emissions from biomass burning have increased in recent decades, largely as a result of increasing rates of deforestation in the tropics. The burning of grasslands, savannas, and agricultural lands has increased over the last century because rarely burned ecosystems, such as forests, have been converted to frequently burned ecosystems, such as grasslands, savannas, and agricultural lands. In Latin America, the area of grasslands, pastures, and agricultural lands increased by about 50% between 1850 and 1985. The same trend is true for South and South-east Asia. Most of the increase results from the ever-increasing rates of forest burning, with other contributions of burning (grasslands, savannas, and agricultural lands) having increased by 15-40%.

Biomass burning has increased with time and that the production of greenhouse gases from biomass burning has increased with time. Furthermore, the bulk of biomass burning is human-initiated. As greenhouse gases build up in the atmosphere and the Earth becomes warmer, there may be an enhanced frequency of fires. The enhanced frequency of fires may prove to be an important positive feedback on a warming Earth.

Cremation of the Dead

The word cremation comes from Latin "cremo," meaning "to burn"—in particular to burn the dead. Temperatures in a crematorium range from about 1400-2100 degrees F. Under these conditions, body and container are almost completely consumed. Remains consist of bone fragments and particles, weighing a total of 4-8 pounds; these are then finely ground. The entire process takes somewhere between 1 and 5 hours. A small portion of the remains is incidentally left inside the cremation chamber and subsequently mingled with the next body to be cremated. Cremation has been practiced for millennia. Archaeologists believe it was invented during the Stone Age, about 3000 B.C. It became the most common method of disposing of bodies in ancient Greece and Rome, and was introduced to Western world by the Greeks as early as 1000 B.C. Greeks had first used it as a way of providing burial for soldiers—those slain in foreign lands were incinerated on the battlefield; their ashes were then gathered and sent home for funerals attended by family and fellow citizens.

The early Christian church initially rejected the practice because of its association with the Pagan societies of Greece and Rome. When Christianity became the official religion of the Roman Empire, and the followers of other religions were exiled or exterminated, burial became the only method of disposing of bodies throughout Europe. Christianity did not expressly forbid cremation, but there was concern that cremation would prevent resurrection of the body. Another worry was a shortage of wood—cremation, as done then, required a lot of timber for pyres. Still, cremation found favor in emergencies. During an outbreak of Black Death in 1656, 60,000 bodies were burned in Naples in a single week.

The first crematorium in the U.S. was built in 1876. In the late 19th century, worry about the unsanitary conditions caused by overflowing cemeteries brought about a revival of cremation. Physicians and sanitary engineers contended that burial in cemeteries saturated the ground with poison. In 1886, the Roman Catholic Church officially banned cremation. As recently as World War II, church members were excommunicated for arranging cremation. Although cremation has now been done in America for about 125 years, it did not really take root here until about 25 years ago. Now about 1 in 4 Americans chooses cremation over traditional burial. Cremation gains favor as society becomes more secular. The trend is fueled by changes in not only religious thought, but also social values and even scientific knowledge. The practice has become common in Europe and Asia faster than in the U.S. It is now almost universal in Japan. The Roman Catholic Church has announced that cremation is not prohibited. The Orthodox Jewish religion, however, continues to forbid it. Modern cremations do not use open fires or copious quantities of firewood.

Cremation can be an aid to murderers. A cremated body cannot be exhumed and analyzed for evidence. There are cases in which a suspicious death occurs, a certain member of the family insists upon cremation. It usually turns out that this one family member is a suspect in the death. It is worth noting that cremation produces carcinogenic dioxins, trace metals, hydrochloric and hydrofluoric acids, sulfur dioxide, and carbon dioxide.

In the Bible, burning of bodies and objects in ancient Israel was reserved mostly for idols and evil material, criminals, the disobedient, and enemies. The New Testament contains few references to burning of bodies, yet fire is used often as a description of hell. Over the history of the church, burning was usually reserved for witches and other heretics. While well-meaning people may have been overzealous in this practice, it is historically significant that fire was reserved for undesirables.

"Cemetery" has its roots in the ancient Greek for "sleeping place," a reflection on the belief that death is a type of sleep. The deceased is buried in a casket and left to decompose over time. Some cultures believe that a body must be whole in the afterlife, which rules out cremation. American traditions try to maintain some connection between those still living and those who have gone before. Cemeteries, especially in the South, are often places of recreation. Reverence for tradition, family, and community supports traditional burial. Burial of the dead is an ecological way of recycling the organic matter back to soil for use by the microbes and plant roots; in the process preventing the release of carbon dioxide gas into the atmosphere which unavoidably occurs in traditional cremation.

The average Indian may go through an entire life without contributing a huge amount to the world's production of greenhouse gases, but in death his carbon footprint jumps. Hindus believe that burning the body entirely helps to release the soul in a cycle of reincarnation that ends only with salvation. But, all the ashes go into the rivers and carbon dioxide is creating global warming. In India, 85 per cent of the billion plus population are Hindus who practice cremation. About ten million people die a year in India. A formal Hindu cremation in which a dead body is burned for more than six hours in a 1m-high open-air pyre can consume more than 400kg of wood to reduce the body to ashes. In case of elite or rich family background, precious wood like sandal and other expensive wood is used to burn the dead body contributing to the release of much more carbon dioxide. The practice of cremation leads to the felling of an estimated 50 million trees, leaves behind half a million tons of ash and produces 8 million tons of carbon dioxide each year.

Urbanization and Industrialization

Modernization is a process of transformation of the way of life (culture, social and economic structures, and attitudes) from the characteristics of traditional societies to those dictated by changes brought about by industrialization, urbanization, trade, and communications. The driving forces for modernization include excessive population growth; mal-distribution of consumption and investment; misuse of technology; corruption and mismanagement; and powerlessness of the victims. The human frailties such as greed, selfishness, intolerance, and shortsightedness; and ignorance, stupidity, apathy, and denial are miscellaneous issues for modernization. These driving forces have linkage with consumption-orientation lifestyles and in effect contribute to the production of greenhouse gases.

Modernization is reflected in urbanization and industrialization. Urbanization means the growth of cities. Around 3 billion or half the world's human population now lives in cities, and virtually all future population growth will take place in cities. Growth of cities means increased pollution as a result of urban traffic, energy consumption and waste production; land taken out of food production, as more and more farmland is built on; loss of natural habitats, as cities and roads proliferate; and fragmentation of rural communities and cultures, as more and more people leave to live in towns. Industrialization is closely linked with population growth and the rise of cities has been a worldwide rapid development of industries such as manufacturing, mining, power generation and transport. These have a major impact on the environment, because they use up huge amounts of water and energy (usually non-renewable fossil-fuel energy) and they are a major agent of climate change, because they release greenhouse gases into the atmosphere and speed up global warming. The built environment is always shaped by climate conditions. The use of new architectural and urban forms, new materials and innovations like air conditioning have driven up both energy costs and the cities' contributions to greenhouse gas emissions.

The "urban heat island effect", resulting from the impacts of land uses in urban areas, can create microclimates and health consequences. Human activities involving the combustion of fossil fuels for industrial or domestic usage, and biomass burning, produce greenhouse gases and aerosols which affect the composition of the atmosphere. The emission of chlorofluorocarbons and other chlorine and bromine compounds has not only an impact on the radiative forcing, but has also led to the depletion of the stratospheric ozone layer. Land-use change, due to urbanization and human forestry and agricultural practices, affect the physical and biological properties of the Earth's surface. Such effects change the radiative forcing and have a potential impact on regional and global climate.

Historically, cities have been the driving force in economic and social development. Urbanization is associated with higher incomes, improved health, higher literacy, improved quality of life and other benefits. These benefits, however, are accompanied by environmental and social ills. These include a diversity of problems ranging from lack of access to clean drinking water to urban air pollution and greenhouse gas emissions. Greenhouse gases are generated from decomposition of organic wastes in landfills and the untreated leachate pollutes surrounding soil and water bodies. Vehicles account for 14 per cent of the world's emissions of carbon dioxide - the primary greenhouse gas in the global warming threat. Nitrogen oxides and hydrocarbons act together to form ozone, also a greenhouse gas. The presence of another vehicular emission, carbon monoxide, can lead to a higher ambient concentration of methane which is another potent greenhouse gas. Finally, chlorofluorocarbons, leaked from air conditioners of vehicles, are not only potent greenhouse gas, but also the main known cause of atmospheric ozone depletion.

For about a thousand years before the Industrial Revolution, the amount of greenhouse gases in the atmosphere remained relatively constant. Since then, the concentration of various greenhouse gases has increased. The amount of carbon dioxide, for example, has increased by more than 30% since pre-industrial times and is still increasing at an unprecedented rate of on average 0.4% per year, mainly due to the combustion of fossil fuels and deforestation. The concentration of other natural radiatively active atmospheric components, such as methane and nitrous oxide, is increasing as well due to agricultural, industrial and other activities. The concentration of the nitrogen oxides and of carbon monoxide is also increasing. Although these gases are not greenhouse gases, they play a role in the atmospheric chemistry and have led to an increase in tropospheric ozone, a greenhouse gas, by 40% since pre-industrial times. Moreover, NO_2 is an important absorber of visible solar radiation. Chlorofluorocarbons and some other halogen compounds do not occur naturally in the atmosphere but have been introduced by human activities. Beside their depleting effect on the stratospheric ozone layer, they are strong greenhouse gases. Their greenhouse effect is only partly compensated for by the depletion of the ozone layer which causes a negative forcing of the surface-troposphere system. All these gases, except tropospheric ozone and its precursors, have long to very long atmospheric lifetimes and therefore become well-mixed throughout the atmosphere.

Human industrial, energy related, and land-use activities also increase the amount of aerosol in the atmosphere, in the form of mineral dust, sulphates and nitrates and soot. Their atmospheric lifetime is short because they are removed by rain. As a result their concentrations are highest near their sources and vary substantially regionally, with global consequences. The increases in greenhouse gas concentrations and aerosol content in the atmosphere result in a change in the radiative forcing to which the climate system must act to restore the radiative balance.

Montreal Protocol

This Protocol on substances that deplete the ozone layer is an international treaty designed to protect the ozone layer by phasing out the production of a number of substances believed to be responsible for ozone depletion. The treaty was opened for signature on September 16, 1987 and entered into force on January 1, 1989 followed by a first meeting in Helsinki, May 1989. It contains provisions for regular review of the adequacy of control measures that are based on assessments of evolving scientific, environmental, technical, and economic information. At a meeting in London in 1990, the Parties to the Montreal Protocol agreed to a phase out of controlled substances. Another meeting of the Parties held in Copenhagen in 1992 accelerated the phase-out schedules of the controlled substances. Controlled substances include CFCs, halons, carbon tetrachloride, methyl chloroform, HCFCs, HBFCs, and methyl bromide. Ozone depleting substances are man-made chemicals that when released into the atmosphere, can cause the depletion of the Ozone layer. This leads to increased levels of ultraviolet radiation reaching the earth's surface, which results in higher rates of skin cancer, eye cataracts and damage to people's immune systems. It also diminishes the productivity of food crops and reduces levels of plankton in the ocean.

Kyoto Protocol

The Kyoto Protocol is a protocol to the international Framework Convention on Climate Change with the objective of reducing greenhouse gases that cause climate change. It was agreed on 11 December 1997 at the 3rd Conference of the Parties to the treaty when they met in Kyoto, and entered into force on 16 February 2005. The Protocol was finally ratified by 146 countries and came into effect in January 2005. An average 5% cut of greenhouse gases emissions has been fixed within 2012, compared to 1990 emissions level. Each country took on a different percentage target within this overall target. An average 5% cut would be a great result (many countries should reduce their emissions instead of increasing them sharply), but unfortunately not enough to stop the temperature increase; for this reason the term "climate change mitigation" is often used. The Kyoto Protocol is an agreement made under the United Nations Framework Convention on Climate Change. Countries that ratify this protocol commit to reducing their emissions of carbon dioxide and five other greenhouse gases, or engaging in emissions trading if they maintain or increase emissions of these gases. Six greenhouse gases are mentioned in the agreement: carbon dioxide, methane, nitrogen protoxide, perfluorocarbons, hydrofluorocarbons and sulfur hexafluoride. Kyoto Protocol does not mention chlorofluorocarbons because they are already regulated in the Montreal Protocol.

In order to get the targets agreed in Kyoto, it is possible to act on gas emissions at local, national or transnational level. Three instruments are foreseen in the Protocol:

1. Emission trading: Forests planted since 1990 are counted as carbon sinks and as credits that offset required cuts in emissions. Also renewable energy projects (solar or wind), or improvement to existing energy generation can be considered.
2. Industrialized countries can buy and sell emission permits, in order to find the cheapest way to cut emissions. The European Union Emission Trading Scheme is the largest multi-national, greenhouse gas emissions trading scheme in the world. It commenced operation in January 2005 and all 25-member states of the European Union participate in the scheme.
3. Joint Implementation and Clean Development Mechanism: They aim at maximizing the cost-effectiveness of climate change mitigation of industrialized countries by allowing them to take up opportunities to cut greenhouse gas emissions by investing in sustainable development projects in foreign countries. Clean Development Mechanism projects allow industrialized countries to fund projects that reduce emissions in developing countries instead of reducing pollution at home.

According to the United Nations Climate Change Secretariat, the Kyoto Protocol's clean development mechanism (CDM) is estimated to generate more than one billion tons of emissions reduction by the end of 2012. The Kyoto Protocol is making a significant contribution towards sustainable development in developing countries.

The Kyoto Protocol requires the industrialized countries to reduce their emissions of carbon dioxide and other greenhouse gases. India ratified the protocol in 2002. The industrialized countries would need to accelerate the introduction of renewables to meet these requirements, as well as reduce energy consumption by promoting efficiency of energy end use, and enhance the absorption of carbon dioxide through greater afforestation. Under the protocol, the developed countries are required to reduce emissions of greenhouse gases by an average of 5.2 per cent below 1990 levels by 2012.

This protocol was adopted in 1997 by the countries that ratified the United Nations Framework Convention on Climate Change (UNFCC). The Convention seeks to stabilize greenhouse gas concentrations in the atmosphere at a level that would minimize interference with the climate system. One of the provisions of the Kyoto Protocol, known as the Clean Development Mechanism (CDM), establishes a framework within which the industrialized countries can meet a part of their carbon dioxide emissions reduction requirements by purchasing Certified Emission Reductions (CERs) from India and other developing countries. These CERs are generated through an international process that continuously monitors and verifies that clean energy projects in developing countries are indeed leading to lower carbon dioxide emissions than would occur otherwise.

The protocol provides for "flexible mechanisms" - ways for countries to reach their targets without actually reducing emissions at home. These include emissions trading - where one country buys the right to emit from another country which has already reduced its emissions sufficiently and has "spare" emissions reductions. Another is the "clean development mechanism" where developed countries can earn credits to offset against their targets by funding clean technologies, such as solar power, in poorer countries. Countries can also claim credits for planting trees that soak up CO_2 - so-called carbon "sinks".

The Kyoto Protocol is therefore an effective international tool to regulate and reduce emissions of greenhouse gases within the permissible levels in order to stabilize the climate and sustain livable environment for all the existing species.

The future of the Kyoto Protocol on climate change is largely in the hands of the world's biggest contributors to greenhouse gas emissions. The US emits more, absolutely and per head, than any other country - although it also produces more wealth. When Kyoto was agreed, the US signed and committed to reducing its emissions by 6%. But since then it has pulled out of the agreement and its carbon dioxide emissions have increased to more than 15% above 1990 levels. For the agreement to become a legally binding treaty, it had to be ratified by countries which together were responsible for at least 55% of the total 1990 emissions reported by the industrialized countries and emerging economies which made commitments to reduce their emissions under the protocol. As the US accounted for 36.1% of those emissions, this 55% target was much harder to achieve without its participation.

All 15 European Union states ratified the Kyoto deal in May 2002. The protocol's most enthusiastic supporter, the EU has pressured countries such as Russia, Japan and Canada to ratify Kyoto so that it could come into force without the commitment of the US. The EU has continually argued for a rigorous application of Kyoto, wanting to limit the use of so-called "flexibility mechanisms" which allow countries to partially meet their emissions reduction targets by paying for improvements in other countries. The EU has also opposed widespread use of forests and other carbon "sinks" to absorb pollution. However, despite its tough stance on Kyoto, the EU is some way off its own target. It pledged to bring total greenhouse gas emissions to 8% below 1990s levels by 2008-2012, but by 2002 they had dropped only 2.9% - and CO_2 emissions had risen slightly. Only four EU countries are on track to achieve their own targets.

China is the world's second biggest emitter of greenhouse gases, but as a developing country is not yet required to reduce its emissions. With China accounting for a fifth of the world's population, increases in its emissions could dwarf any cuts made by the industrialized countries. The average Chinese person consumes only 10-15% of the energy an average US citizen uses, but with the economy developing at high speed many analysts expect China's total emissions to overtake America's by mid-century. Fossil fuels play a major role - China is the world's biggest coal producer and oil consumption has doubled in the last 20 years. The country faced power cuts in 2004 as soaring growth outstripped electricity generation. China's leaders recognize that climate change could devastate their society and ratified the Kyoto Protocol in 2002. In 2004 Beijing announced plans to generate 10% of its power from renewable sources by 2010. But it is far from clear whether the country would ever agree to internationally-imposed emissions restrictions.

Russia ratified the Kyoto Protocol in November 2004 - the crucial moment making the treaty legally binding. Russia's entry was vital, because the protocol had to be ratified by nations accounting for at least 55% of greenhouse gas emissions to become valid. This target was only met after Russia joined. Russia's economy has shrunk so drastically since 1990 that industrial activity has dropped, leaving emissions reduced by about 35% and well below the level allowed under Kyoto. In the short-term, Russia stands to gain billions of dollars through emissions trading - selling its unused emissions entitlement to developed countries which want to emit more than the protocol allows them to. The money would be used for energy efficiency projects. Committing to keep emissions low could, however, bring Russia economic costs in the longer term.

A major world economic power, Japan is a leading member of Kyoto, committed to cutting emissions. It was responsible for 8.5% of emissions in 1990 and its support for the agreement has been critical in the absence of US participation. Although previously reluctant to ratify the protocol unless the US also committed, Japan ratified it in June 2002. It committed to reduce emissions by 6% from 1990 levels, but 2002 figures showed total greenhouse gas emissions had risen 11% above the baseline figure. The country recognizes that its economy could benefit from the Kyoto agreement, as Japanese companies could capture markets for new, clean technology.

Developing countries like India are not obliged to make any cuts in greenhouse emissions under Kyoto. But as they raise living standards their emissions will increase. India's emissions are estimated to have risen by more than 50% in the 1990s, although the country has only submitted emissions figures to the UN for one year, 1994. India recognizes that many of its one billion people will be vulnerable to the effects of climate change and ratified the Kyoto Protocol in August 2002. But, with India's economy and population, like China's, continuing to grow, the thorny issue of developing country emissions commitments will have to be tackled soon in future rounds of negotiations.

In developing countries which are reeling under fragile economics, the climate change mitigation will bring additional strain and will affect the efforts to achieve higher economic growth rates to eradicate poverty speedily. These countries do not have adequate resources to meet their human needs. The mitigation measures to reverse global warming effects simply mean reversing the growing economies which are based on consumption of materials. But, if climate change mitigation is neglected, it would also take its own toll in different forms such as food crisis, energy crisis, land use changes and eventually all these collectively affecting the economies of these countries. Therefore, a balanced approach is required for the developing countries to deal with the effects of global warming while progressing well with growing trend of economies.

2

Global Warming

The atmosphere's temperature decreases with height in the troposphere. Most of the radiation emitted from the upper atmosphere escapes to space, while most of the radiation emitted from the lower atmosphere is re-absorbed by the surface or the atmosphere. Thus, the strength of the greenhouse effect depends on the atmosphere's rate of temperature decrease with height: if the rate of temperature decrease is greater the greenhouse effect will be stronger, and if the rate of temperature decrease is smaller then the greenhouse effect will be weaker.

Global warming, a recent warming of the Earth's lower atmosphere, is believed to be the result of an enhanced greenhouse effect which is due to increased concentrations of greenhouse gases in the atmosphere. In addition to the Earth, Mars and Venus have greenhouse effects. Global warming is the sustained increase in the average temperature of the Earth's near-surface air and oceans since the mid-twentieth century and its projected continuation. The average global air temperature near the Earth's surface increased 0.74 ± 0.18 °C during the hundred years ending in 2005. Most of the observed increase in globally averaged temperatures since the mid-twentieth century is attributed to the observed increase in anthropogenic greenhouse gas concentrations via the greenhouse effect. Global temperatures are expected to continue to rise as human activities continue to add carbon dioxide, methane, nitrous oxide, and other greenhouse gases to the atmosphere. The extent of climate change effects, and whether these effects prove harmful or beneficial, will vary by region, over time, and with the ability of different societal and environmental systems to adapt to or cope with the change.

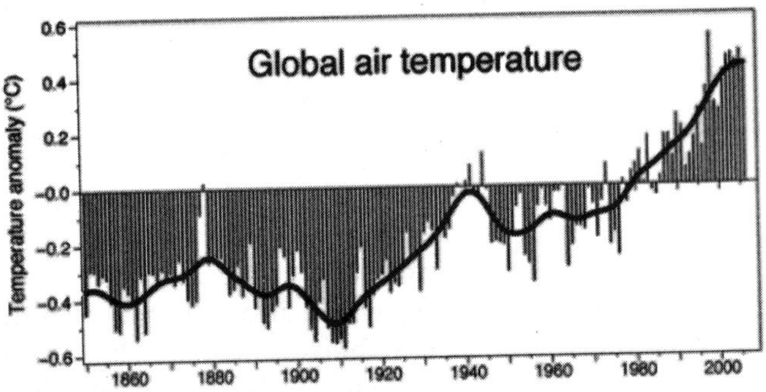

India's carbon dioxide equivalent emissions of greenhouse gases could increase to 3,000 million tons by 2020 which would be about twice the emission level in 2000. Even so, India's emissions would account for less than five per cent of global emissions in 2020 and its per capita emissions would be low when compared to most developed countries. Since greenhouse gas emissions were directly linked to economic growth, India's economic activities would necessarily involve increased emissions. Any constraint on emission would hamper economic development.

Food Production

Global temperatures will rise by an average of 3°C due to climate change and cause catastrophic damage around the world. It would cause a worldwide drop in cereal crops between 20 and 400 million tons, put 400 million more people at risk of hunger and put up to 3 billion people at risk of flooding and without access to fresh water supplies. Few ecosystems could adapt to such a temperature change equivalent to a level of carbon dioxide of 550 parts per million in the atmosphere which would result in the destruction of half the world's nature reserves and a fifth of coastal wetlands.

The effects of climate change begin to bite in terms of empty shelves in Caracas, food riots in Mexico, warnings of hunger in Jamaica, Nepal, the Philippines, and sub-Saharan Africa. In other developing countries, food shortage and price hike in food and food products begin to appear due to failure or inadequate amount of rainfall. Soaring prices for basic foods are beginning to lead to political instability with governments being forced to step in to artificially control the cost of bread, maize, rice and dairy products.

Record world prices for most staple foods have led to 18 per cent food price inflation in China, 13 per cent in Indonesia and Pakistan and 10 per cent or more in Latin America and Russia, according to the United Nations Food and Agricultural Organization (FAO). Wheat has doubled in price, maize is nearly 50 per cent higher than a year ago and rice is 20 per cent more expensive according to United Nations. FAO is expected to say that global food reserves are at their lowest in 25 years and those prices will remain high for years. In Russia, there are signs for food companies to freeze the price of milk, bread, and other foods for fear of a public backlash. Yemen, Mexico and several other countries have had, or been close to food riots in the present times, something not seen in decades of low global food commodity prices. Meanwhile, there are shortages of beef, chicken and milk in Venezuela and other countries as governments try to keep a lid on food price inflation. Boycotts have become commonplace. Argentineans shunned tomatoes during the recent presidential election campaign when they became more expensive than meat. Italians organized boycotts of pasta in protest at rising prices. German left wing politicians have called for an increase in welfare benefits so that people can cope with price rises.

The increase of the oil prices and the increase of food prices put together would show the elements of a very serious social crisis in the future. The price rises are a result of record oil prices. U.S. farmers are switching out of cereals to grow bio-fuel crops under extreme weather. They distorted the world market for cereals by growing 14 million tons, or 20 per cent of the whole maize crop for ethanol for vehicles. This took millions of hectares of land out of food production and nearly doubled the price of maize. Further, some more per cent of crop land is expected to be used for cereal production instead of maize crop. Cereal stocks have been declining for more than a decade but now stood at around 57 days, which made global food supplies vulnerable to an international crisis or big natural disaster such as a drought or flood. Any unforeseen flood or crisis can make prices rise very quickly. According to World Watch Institute, the competition for grain between the world's 800 million motorists who want to maintain their mobility and its 2 billion poorest people who are simply trying to survive is emerging as an epic issue.

Maize is a staple food in many countries which import from the U.S., including Japan, Egypt and Mexico. U.S. exports are 70 per cent of the world total and are used widely from animal feed. The shortages have disrupted livestock and poultry industries worldwide. The policy of European Union that substituting 10 per cent of all car fuel with bio-fuels is expected to threaten the lives of poor farmers. The crisis is being compounded by growing populations, extreme weather and ecological stress according to a number of recent reports. The United Nations Environment Programme disclosed that the planet's water, land, air, plants, animals and fish stocks were all in inexorable decline. According to the United Nations World Food Programme, 57 countries including 29 in Africa, 19 in Asia and 9 in Latin America have been hit by catastrophic floods. Harvests have been affected by drought and heat waves in south Asia, Europe, China, Sudan, Mozambique and Uruguay.

Rising temperatures and varying rainfall patterns could affect staple crop production and food security, while aiding the migration and breeding of pests that can devastate crops. Farmers in the tropical developing world will likely to see decreases in production. Such changes could be devastating to people in poor countries, even while some cold climate nations, such as Canada, may expand their arable land. With the prices of wheat, rice, and other staples already rising rapidly, the developing world can ill afford any production decreases at home. In addition, more severe weather, such as monsoons or hurricanes, can destroy crops and leave entire communities without food.

Poorer countries with meager resources and weak health systems will be hit hardest because malnutrition is already widespread, with the young, women and the elderly at risk. Millions of people could face poverty, disease and hunger. Unexpected climate patterns - too much rain or too little - will have an impact on food production, especially irrigated crops such as rice, and can cause unemployment, economic upheavals and political unrest. There is a burden on the governments to strengthen current systems providing clean water, immunizations, disease surveillance, mosquito control and disaster preparedness.

Global warming is bound to affect food supply and security, especially in developing countries. Even if developing countries adapt to climate change, they will not be able to completely avoid the problems associated with climate change. Furthermore, these harmful outcomes of climate change in developing countries and potentially positive outcomes in developed countries will probably increase the gap in wealth, access to food, and health between rich and poor countries. This will affect the worldwide economy due to loss of emerging trade-partners, increases in hunger, and departure of refugees from their regions in search of food and resources.

Today the food system is even more reliant on cheap crude oil. Virtually all of the processes in the modern food system are now dependent upon this finite resource, which is nearing its depletion phase. The systems that produce the world's food supply are heavily dependent on fossil fuels. Vast amounts of oil and gas are used as raw materials and energy in the manufacture of fertilizers and pesticides, and as cheap and readily available energy at all stages of food production: from planting, irrigation, feeding and harvesting, through to processing, distribution and packaging. In addition, fossil fuels are essential in the construction and the repair of equipment and infrastructure needed to facilitate this industry, including farm machinery, processing facilities, storage, ships, trucks and roads. The industrial food supply system is one of the biggest consumers of fossil fuels and one of the greatest producers of greenhouse gases.

Ironically, the food industry is at serious risk from global warming caused by these greenhouse gases, through the disruption of the predictable climatic cycles on which agriculture depends. But global warming can have the more pronounced and immediate effect of exacerbating existing environmental threats to agriculture, many of which are caused by industrial agriculture itself. Environmental degradation, water shortages, salination, soil erosion, pests, disease and desertification all pose serious threats to our food supply, and are made worse by climate change. But many of the conventional ways used to overcome these environmental problems further increase the consumption of finite oil and gas reserves. Thus the cycle of oil dependence and environmental degradation continues.

Industrial agriculture and the systems of food supply are also responsible for the erosion of communities throughout the world. This social degradation is compounded by trade rules and policies, by the profit driven mindset of the industry, and by the lack of knowledge of the faults of the current systems and the possibilities of alternatives. But the globalization and corporate control that seriously threaten society and the stability of our environment are only possible because cheap energy is used to replace labour and allows the distance between producer and consumer to be extended.

However, this is set to change. Oil output is expected to peak in the next few years and steadily decline thereafter. We have a very poor understanding of how the extreme fluctuations in the availability and cost of both oil and natural gas will affect the global food supply systems, and how they will be able to adapt to the decreasing availability of energy. In the near future, environmental threats will combine with energy scarcity to cause significant food shortages and sharp increases in prices - at the very least. The humanity is about to enter an era where it will have to once again feed the world with limited use of fossil fuels. But the point is that do we have enough time, knowledge, money, energy and political power to make this massive transformation to our food systems when they are already threatened by significant environmental stresses and increasing corporate control?

Food Insecurity in India

Rising food prices could spark worldwide unrest and threaten political stability. Food scarcity and soaring fuel prices would compound the damaging effects of global warming. Food riots are already being reported across the globe. Current food price trends are likely to increase sharply both the incidence and depth of food insecurity.

Providing food security to the urban population has been made more vital by the fact that urbanization has been a dominating trend throughout the world in the last half century. According to the 1961 Census of India, 17.97 per cent of the country's population lived in urban areas. In 2001, that figure was 27.78 per cent, amounting to 285 million people. The United Nations estimates that India's urban population will reach 600 million by the year 2025. Urbanization has already eaten up vast quantities of productive land and has placed enormous pressure on the available infrastructure. The past decade of liberalization has brought about vast changes in the country, the effects of which have been more pronounced in urban areas. Yet, the specter of urban poverty haunts India's crowded cities, and providing food security to the urban population has and will continue to be a tremendous challenge.

On the surface, life for all sections in urban India appears to be easier than in rural India. Wages and salaries are higher in urban areas; infrastructure is superior in cities and towns when compared with villages; schools and hospitals are more accessible; food availability is rarely a problem; and small signs of wealth such as radios and televisions are common, even in the slums. However, such observations are deceptive. Livelihood access is vital in achieving urban food security. Large sections of lower income groups depend on casual employment or are self-employed in petty businesses and these types of employment are usually accompanied by uncertain incomes. For the country as a whole, more than 14 per cent of the urban population is dependent on casual labour as a means of livelihood. For the lowest 10 per cent of the urban population in India, 37.49 per cent are engaged in casual labour and 41.34 per cent are self-employed, suggesting that a vast majority of the urban poor are vulnerable to uncertain incomes and, hence, vulnerable to under-nourishment. The uncertainty of these avenues of income has a significant effect on the food security of the urban poor.

The nineties ended with a baffling accumulation of food stocks in the midst of widespread hunger. Food grain procurement leaped from an average of 20 million tons per year in 1991-1997 to 25 million tons in 1998, 30 million tons in 1999, 35 million tons in 2000 and 40 million tons in 2001. Meanwhile, food grain off-take through the public distribution system did not increase. In fact, it declined from 17 million tons per year in the early nineties to 12 million tons per year around the end of the decade. The result was a yawning gap between procurement and distribution, and ballooning food stocks. India's food grain mountain, the largest in world history by a long margin, reached a peak of nearly 70 million tons in mid-2002.

The most incongruous part of the story is that the bulk of this stock was accumulated during a period of intense drought in large parts of the country. At a time of widespread hunger in drought-affected areas, the government was busy in hoarding food on an unprecedented scale and striving to keep prices up, making it much harder for drought-affected people to buy food from the market. The availability of massive food stocks during the last few years have not led to any major initiative in the field of food security. The silver lining is that this situation has led to greater awareness of the problem of endemic hunger and opened up new possibilities of public action. The monstrosity of the situation that prevailed during the last few years, as people starved in the shadow of food mountains has led to a healthy renewal of public concern with the issue. The coverage of hunger and nutrition matters in the print media, for instance, has sharply increased. And perhaps for the first time in independent India, there are good prospects of endemic hunger becoming a lively political issue.

Recent years have seen the economy booming and growth rates have been among the highest in the world. The flip side however, is that one in every five Indians suffers from overt or covert hunger. India is an example of how food sufficiency at the aggregate level has not translated into food security at the household level. A staggeringly large number of undernourished about 214 million people is chronically food insecure. Many more varyingly about 40 million are exposed to natural disasters. About 50 per cent of children (mostly tribal and rural) are undernourished and stunted, 23 per cent have a low birth weight and 68 out of 100 die before the age of one year. There is a high prevalence of anemia and other micronutrient deficiencies.

Air Pollution

It is the addition of harmful substances to the atmosphere resulting in damage to the environment, human health, and quality of life. Air pollution makes people sick—it causes breathing problems and promotes cancer—and it harms plants, animals, and the ecosystems in which they live. Some air pollutants return to Earth in the form of acid rain and snow, which corrode statues and buildings, damage crops and forests, and make lakes and streams unsuitable for fish and other plant and animal life. Pollution is changing Earth's atmosphere so that it lets in more harmful radiation from the Sun. At the same time, the polluted atmosphere is becoming a better insulator, preventing heat from escaping back into space and leading to a rise in global average temperatures.

Most air pollution comes from one human activity: burning

fossil fuels—natural gas, coal, and oil—to power industrial processes and motor vehicles. Among the harmful chemical compounds this burning puts into the atmosphere are carbon dioxide, carbon monoxide, nitrogen oxides, sulfur dioxide, and tiny solid particles—including lead from gasoline additives—called particulates. Some of these pollutants also come from natural sources. Forest fires emit particulates and volatile organic chemicals into the atmosphere. Volcanoes spew out sulfur dioxide and large amounts of pulverized lava rock known as volcanic ash. A big volcanic eruption can darken the sky over a wide region and affect the Earth's entire atmosphere. The 1991 eruption of Mount Pinatubo in the Philippines dumped enough volcanic ash into the upper atmosphere which in turn lowered global temperatures for the next two years. Unlike pollutants from human activity, however, naturally occurring pollutants tend to remain in the atmosphere for a short time and do not lead to permanent atmospheric change. Once in the atmosphere, pollutants often undergo chemical reactions that produce additional harmful compounds. Air pollution is subject to weather patterns that can trap it in valleys or blow it across the globe to damage pristine environments far from the original sources.

Local and regional pollution take place in the lowest layer of the atmosphere, the troposphere, which at its widest extends from Earth's surface to about 16 km. The troposphere is the region in which most weather occurs. If the load of pollutants added to the troposphere were equally distributed, the pollutants would be spread over vast areas and the air pollution might almost escape our notice. Pollution sources tend to be concentrated, however, especially in cities. In the weather phenomenon known as thermal inversion, a layer of cooler air is trapped near the ground by a layer of warmer air above. When this occurs, normal air mixing almost ceases and pollutants are trapped in the lower layer. Local topography, or the shape of the land, can worsen this effect—an area ringed by mountains can become a pollution trap.

Smog forms when sunlight, heat and relatively stagnant air meet up with nitrogen oxides and various volatile organic compounds. It is intense local pollution usually trapped by a thermal inversion. Before the age of the automobile, most smog came from burning coal. In 19th-century London, smog was so severe that street lights were turned on by noon because soot and smog darkened the midday sky. Burning gasoline in motor vehicles is the main source of smog in most regions today. Powered by sunlight, oxides of nitrogen and volatile organic compounds react in the atmosphere to produce photochemical smog. Smog contains ozone which if present in the lower atmosphere is a poison. It damages vegetation, kills trees, irritates lung tissues, and attacks rubber. Exposure to smog can do serious damage to our lungs and respiratory systems. Inflammation and irritation can cause shortness of breath, throat irritation, chest pains and coughing and lead to asthma attacks, hospital admissions and emergency room visits. These consequences are more severe if people are exposed while being active. More hot days mean better conditions for creating smog that can trigger asthma and other breathing problems.

In the presence of atmospheric moisture, sulfur dioxide and oxides of nitrogen turn into droplets of pure acid floating in smog. These airborne acids are bad for the lungs and attack anything made of limestone, marble, or metal. Smog acids are eroding precious artifacts like the Taj Mahal in Âgra, India. Oxides of nitrogen and sulfur dioxide pollute places far from the points where they are released into the air. Carried by winds in the troposphere, they can reach distant regions where they descend in acid form, usually as rain or snow. Such acid precipitation can burn the leaves of plants and make lakes too acidic to support fish and other living things. Smog spoils views and makes outdoor activity unpleasant. For the very young, the very old, and the people who suffer from asthma or heart disease, the effects of smog are even worse: It may cause headaches or dizziness and can cause breathing difficulties. In extreme cases, smog can lead to mass illness and death, mainly from carbon monoxide poisoning.

With stronger pollution controls and less reliance on coal for heat, today's chronic smog is rarely so obviously deadly.

However, under adverse weather conditions, accidental releases of toxic substances can be equally disastrous. The worst such accident occurred in 1984 in Bhopal, India, when methyl isocyanate released from an American-owned factory during a thermal inversion caused more than 3,800 deaths.

Air pollution worsens as temperatures increase, and higher levels of ozone, smog and other pollutants have been directly linked with increased rates of respiratory and cardiovascular disease, including asthma and cardiac problems. Pediatric asthma has already increased over the past twenty five years, and global warming will only exacerbate children's suffering.

Wildfire

Severe heat can also increase the frequency and intensity of wildfires, which threaten homes, lives, and livelihoods, and cause poor air quality. A warming climate encourages wildfires through a longer summer period that dries fuels, promoting easier ignition and faster spread of fire. The result is serious loss of life and property. The fires could reduce the diversity of ecosystems and species diversity. Further, the positive and negative interactions between different species could also be disrupted by fires. Certain plant species would have a better chance of survival and establishment. The fires also pose air pollution problems as biomass burning produces toxic volatiles, soot, particulate matter, carbon dioxide and other chemicals which contribute to global warming and change in atmospheric chemistry locally.

In the United States of America, year-to-year changes in wildfire frequency have been shown to be strongly linked to annual spring and summer temperatures and to the timing of spring snowmelt. At higher elevations what really drives the fire season is the temperature. With the snowmelt coming out a month earlier, areas get drier overall. There is a longer season in which a fire can be started and more opportunity for ignition. In recent years, wildfires in the western United States have burned hundreds of homes annually and caused extreme and sometimes irreversible damage to natural resources. Fire-fighting expenditures for wildfires now regularly exceed one billion dollars per year. More severe fires could change forest composition so drastically and the forests which currently store atmospheric carbon dioxide could become a source of atmospheric carbon dioxide. Additional carbon dioxide could further warm the climate and exacerbate the fire problem.

Similarly, in other parts of the world, there are frequent reports pouring in stating that wildfires have become very frequent and causing devastating effects on forests. In effect, the forest species composition, landscape and soil constituents all are being altered. The wildfires are converting the forests which are the storehouses of carbon into vacant space previously occupied forests and pumping carbon dioxide into the atmosphere. The carbon dioxide levels thus released are amplifying the effects of global warming.

Oceans and Sea Level Rise

Increased atmospheric carbon dioxide increases the amount of carbon dioxide dissolved in the oceans. Carbon dioxide dissolved in the ocean reacts with water to form carbonic acid, resulting in acidification. Ocean surface pH is estimated to have decreased from 8.25 near the beginning of the industrial era to 8.14 by 2004, and is projected to decrease by a further 0.14 to 0.5 units by 2100 as the ocean absorbs more carbon dioxide. Since organisms and ecosystems are adapted to a narrow range of pH, this raises extinction concerns, directly driven by increased atmospheric carbon dioxide, that could disrupt food webs and impact human societies that depend on marine ecosystem services.

When global temperatures increase, ice near the poles melts at an increasing rate. As the ice melts, land or open water takes its place. Both land and open water are on average less reflective than ice, and thus absorb more solar radiation. This causes more warming, which in turn causes more melting, and this cycle continues. This contributes to a rise in sea levels. Sea levels could rise between 9 and 88 centimeters by the year 2100. A rise of this magnitude would have disastrous consequences for people living on low-lying islands, such as the Maldives group in the Indian Ocean and many South Pacific islands. Rising sea levels make coastal areas more susceptible to storm surges and flooding that result from severe weather. The most susceptible areas are densely populated river deltas and coastal cities in Asia—the Ganges River Delta, the Mekong River Delta, islands in the Indian Ocean and South Pacific. With displacement comes increased transmission of water-borne disease from stagnant water, the challenge of feeding and sheltering the displaced, sewage backups and squalid conditions, and strained disaster relief resources. Related problems are the contamination of coastal freshwater supplies with encroaching sea water, and the degradation of fishing and agricultural areas.

Mountains are the cathedrals of the Earth, and glaciers are the barometers of our planet's climate. The pleasure and intense enjoyment humanity got from mountains lives on but the cathedrals are crumbling. Now more than ever mountains matter. They are the ultimate litmus test and the most immediate graphic indicator of what is happening to the Earth's climate. By 1975 that volume of ice had halved. But what is remarkable is the speed of melting. A further 25 per cent of ice melted between 1975 and 2000, and an even faster 10-15 per cent in the first five years of this century.

More serious is the melting of permafrost, the subterranean glue that has cemented apparently solid structures for centuries. The shattering of our playground would not matter if it were not such a clear demonstration of bigger concerns both the alarming speed of the current warming worldwide and the specific threat to mountain communities threatened by destabilized mountain slopes. Of course mountain landscapes particularly young mountain ranges such as the Alps, Himalayas and Andes are inherently unstable but the instability threatens to become much more acute. In Nepal, the speeding retreat of glaciers is producing ever more menacing melt water lakes. In the case of some rivers such as the Nile about 90 per cent of the flow derives from mountains. The Punjab, the bread bowl of northern India and Pakistan is named after the five rivers that all rise in the Himalayas close to that holy mountain Kailash revered by Hindus and Buddhists. About 1.5 billion people in China and South-East Asia get their water from the Himalayan ranges. The mountains matter not just as a place of escape but as the very source of life.

The World Bank describes Egypt as particularly vulnerable to the effects of global warming because it faces potentially catastrophic consequences. A big reason is vulnerability of Egypt's bread basket - Nile delta, an area of rich, arable land where the Nile River spreads out and drains into the Mediterranean Sea. Although the Delta makes up only 2.5 per cent of Egypt's land mass, it is home to more than a third of this largely desert country's 80 million people. The Delta was already in danger threatened by the side effects of southern Egypt's Aswan Dam. Though the dam, completed in 1970, generates much needed electricity and controls Nile River flooding, it also keeps nutrient sediment from replenishing the eroding Delta. Add climate change to the mix, and the Delta faces new uncertainties that could have a potentially more devastating effect on Egypt. The Mediterranean, and the world's other seas, would rise between one foot (30 centimeters) and 3.3 feet (one meter) by the end of the century, flooding coastal areas along the Delta. A rise of 3.3 feet would flood a quarter of the Delta, forcing about 10.5 per cent of Egypt's population from their homes, according to the World Bank. By 2100 the rising waters could wipe out the sandy beaches that attract thousands of tourists.

The melting of the icebergs would increase the sea level by almost 33 inches in the next few decades and that would completely submerge a few island nations and coastal areas of many other countries. The climatic change would be disastrous and would hit the mankind on all fronts right from water to air and from food security to natural calamities. Further, it would result in the extinction of over 50 per cent of species.

Glacial Red Alert

News from the Antarctic is that while a large part of the icy continent is cooling, other parts are getting warmer. The ozone hole over the South Pole is growing wider exposing us to harmful ultra violet radiation while greenhouse gases continue to corrode the atmosphere.

A recent United Nations Environment Programme report based on a study in Bhutan and Nepal has reported rising water levels in at least 44 glacial lakes as increasing temperatures accelerated the melting of glaciers and surrounding snowfields. Large-scale loss of property and business is feared; the United Nations Environment Programme warns that inhabitants of endangered valleys will need to be relocated to safer places. Indian Himalayan glaciers are faced with similar meltdowns and the government is reportedly setting up a national centre to study the problem in partnership with the DRDO. The implications of this are wide: flooded valleys, acute shortage of clean drinking water, threat from water-borne diseases, adverse effect of silt on hydro-power plants located upstream and relocations of troops stationed in glacial areas.

The Himalayan glaciers have started to melt and the average rate of retreat is almost twice (34 meters) per year as compared to the 1971 levels of 19 meters. The melting glaciers will cause temperatures and sea-levels to rise and there will be a cascading effect on the crops and the monsoons. Rising sea levels have submerged two islands in the Sunder bans, where tigers roam through mangrove forests in the Ganges River delta, and a dozen more islands are under threat. The effects of global warming on India vary from the submergence of low-lying islands and coastal lands to the melting of glaciers in the Indian Himalayas, threatening the volumetric flow rate of many of the most important rivers of India and South Asia. In India, such effects are projected to impact millions of lives. As a result of ongoing climate change, the climate of India has become increasingly volatile over the past several decades; this trend is expected to continue.

The Geological Survey of India's study of Himalayan glaciers shows that the retreat has been accelerating at a higher rate over the past 10 years. Global temperatures have risen by 0.6°C in the last century and they are expected to rise by 1.1°C in the next 25 years. Global warming has to be dealt with on several fronts, the foremost being reducing fossil fuel use, cutting down on poisonous emissions, finding alternative sources of renewable energy, developing cleaner technology, and reforestation and co-opting the services of huge multinationals across the world to clean the air and promote public health. Good disaster management on the other hand has to deal with both preventive steps, anticipatory relief and rehabilitation measures. These would entail long-term damage control in reducing green house gas emissions and creating public awareness while dealing with the problems of providing potable water and medical services. Since the meltdown is global, the effort to deal with it has also got to be global. This means synergizing data and analyses and charting out plans of action for international cooperation and compliance.

Coastal Ecosystems in India

The coastal systems are among the most vulnerable due to the rapidly warming global environment is a well known and understood fact. Indian coastline stretches about 5700 kms on the mainland and about 7500 kms including the two island territories and exhibits most of the known geomorphological features of coastal zones. Presently, Indian coastline is facing increasing human pressures e.g., overexploitation of marine resources, dumping of industrial and toxic wastes, oil spills and leaks which have resulted in substantial damage to its ecosystems. The impact of global warming-induced sea level rise due to thermal expansion of near-surface ocean water has great significance to India due to its extensive low-lying densely populated coastal zone. Sea level rise is likely to result in loss of land due to submergence of coastal areas, inland extension of saline intrusion and ground water contamination and may have wide economic, cultural and ecological repercussions. A mean sea level rise of between 15 and 38 cm is projected by the mid- 21st century along India's coast. Added to this, a 15% projected increase in the intensity of tropical cyclones would significantly enhance the vulnerability of populations living in cyclone prone coastal regions of India. Other sectors vulnerable to the climate change include freshwater resources, industry, agriculture, fisheries, tourism ,and human settlements. Given that many climate change impacts on India's coastal zone feature irreversible effects, the appropriate national policy response should enhance the resilience and adaptation potential of these areas.

India has been identified as one amongst 27 countries which are most vulnerable to the impacts of global warming related to accelerated sea level rise. The high degree of vulnerability of Indian coasts can be mainly attributed to extensive low-lying coastal area, high population density, frequent occurrence of cyclones and storms, high rate of coastal environmental degradation on account of pollution and non-sustainable development. Most of the people residing in coastal zones are directly dependent on natural resource bases of coastal ecosystems. Any global warming-induced climatic change such as increase in sea surface temperature, change in frequency, intensity or tracks of cyclones, sea level rise may aggravate the potential risks to coastal zones. The rise in sea level could result in the loss of cultivable land due to inundation, salt water intrusion into coastal ecosystems and into groundwater systems and loss of terrestrial and marine biodiversity.

Coastal zones in India are currently facing increasing stresses and shocks as a consequence of cumulative environment change driven by population growth, urbanization, industrial development, trade and capital flows, liberalization of transnational corporation activity and life style, attitudinal changes and unsustainable growth leading to degradation of coastal zones and ecosystems. A large percentage of coastal population is directly dependent on coastal and marine resources like mangroves and other coastal vegetations, agriculture and fisheries for their livelihood. Any impact to the coastal ecosystem may affect the coastal population in almost every sector of their lives. Exploration and exploitation of oil and gas activities in the offshore regions like Godawari, Krishna and Bombay have resulted in pollution of coastal ecosystems. The problem of freshwater is already acute in coastal regions of Tamil Nadu, Gujarat, West Bengal and Orissa due to the over-extraction of groundwater which has resulted in lowering of groundwater table and salt water intrusion. The lowering of groundwater table has also led to subsidence of coastal land causing high relative sea level rise. Lagoons located across the coastal regions of India tend to be decreasing in area due to silting of sediments and plant growing. Sea-level rise and reduction of river-borne sediments will decelerate delta progradation and wetland renewal. The rich biodiversity is seriously threatened by loss of wetlands due to sea-level rise.

As a result of major coral bleaching in 1998, many reefs dominated by branching species in Indian Seas have been severely damaged with high mortality of these species. In coastal seas around Andaman Islands of India, reef community structure has switched from dominance by fast-growing branching species to monopolization by the more physically rigorous and slow-growing massive corals. Quarrying of coral reefs in Lakshadweep islands, Gulf of Mannar in Tamil Nadu and Gulf of Kutch in Gujarat for the manufacture of calcium carbonate and construction purposes has exposed the coasts to the eroding actions of the waves. Mining of monazite sands for extracting a number of rare earth elements like ilmenite, zircon and rutile along Kerala coast at levels below sea water has resulted in submergence of land, salt water intrusion into fresh water and erosion. Mining of cast iron ore along Goa coast has led to the formation of acidic waste lands making the land more vulnerable to the impacts of sea-level rise and erosion. The vulnerability of coastal zones to an increasing range of stresses and shocks are likely to be exacerbated by climate change.

Global warming would result in an increase in sea surface

temperatures as a result of which changes in the frequency, intensity or tracks in cyclones hitting the coastal zones may take place. India is frequently affected by tropical cyclones and storms particularly during or around monsoon season. About 6.5% of about world's 80 tropical storms are formed annually in the Indian Ocean. Frequency of formation of cyclones is 5-6 times more in Bay of Bengal as compared to Arabian Sea. Therefore, the east coast of India (particularly the States of Andhra Pradesh, Orissa and West Bengal) is more vulnerable than the west coast to the fury of cyclones and massive damage of life and property occurs almost every year. Storm surges are generated by the winds and the atmospheric pressure changes associated with cyclones. At low latitude land-locked locations such as Bay of Bengal, tropical cyclones are the major cause of storm surges. Amplification in storm surge heights should result from the occurrence of stronger winds and low pressure associated with more intense tropical storms. Thus, an increase in sea surface temperature due to climate change should lead to higher peaks of storm surges and a greater risk of coastal disasters along the east coast of India in the event of sea-level rise.

The potential impact of one meter sea-level rise along Indian coast provides an idea about the land which could be inundated and the population that would be affected provided no protective measures are taken. It has been suggested that the total area of 5763 km^2 along the Coastal States of India i.e., 0.41% could be inundated and almost 7.1 million i.e., 4.6% of coastal population could be directly affected. The most vulnerable areas along the Indian coastline are the Kutch region of Gujarat, Mumbai and South Kerala. Deltas of rivers Ganges (West Bengal), Cauvery (Tamil Nadu), Krishna and Godawari (Andhra Pradesh), Mahanadi (Orissa) and also the islands of Lakshadweep Archipelago would be totally lost.

In the coastal regions of Tamil Nadu, salinity of groundwater due to the intrusion of seawater into the subsurface aquifer is a major problem. Due to excess withdrawal of groundwater, the water table has fallen too far below thereby allowing seawater to percolate. Similarly, in Gujarat due to uncontrolled withdrawal of groundwater, the ground water is becoming highly saline apart from the fact that depth of the water table reaching at places beyond 200 meters. Coastal aquifer system will be more contaminated with salinity bringing greater complicacy to the problem of tapping usable groundwater. In coastal regions of West Bengal and Orissa, the problem of fresh water is fairly acute because of the depth of water table and high cost of lifting the same from the depth of 700-1000 meters. The shallow salt water table often renders stored water in ditches and ponds brackish and surface soil saline. In view of this, only one kharif crop could be possible. Tidal ingress and pushing up of saline waters inland may extend by 35-50 kms beyond the present limit and during storm conditions, the spread of salinity in the low-lying agricultural lands may ruin the prospect of crops.

Intensive production systems and large scale facilities used to raise high value shrimp, salmon and other premium species have taken a heavy toll on coastal habitats, with mangrove swamps in India being cleared at an alarming rate. Coastal ecosystems like mangroves, coral reefs provide natural protection to coasts by dissipating considerable wave energy and hence only smaller waves of moderate intensity reach the shore and break. Coastal erosion is caused by forces of nature but very often accentuated by anthropogenic activities like making of structures on shores, removal of the materials from the shores etc. The fishery resources of India are being depleted by over fishing, excessive use of pesticide, industrial pollution and even construction of coastal structures. The loss of inshore fish nursery habitats by coastal development, and pollution from land-based activities cause significant change to ecosystems supporting fisheries. The impact of global warming on fishery will depend on the complicated food chain, which could be disturbed by sea-level rise, change in ocean currents, and alteration of the mixing layer thickness.

The rich biodiversity of the wetlands in India is seriously threatened by loss of wetlands due to sea-level rise. India's

Sunder bans are highly vulnerable to sea-level rise since it will change the salinity distribution and hence the productivity. The effect of sea-level rise will depend on the type of mangrove forest, these may either keep pace with the rising sea level rise or these may be submerged. Large scale changes in species composition and zonation in mangrove forests are also expected due to changes in sedimentation and organic accumulation, nature of coastal profile and species interaction.

The Sunder bans in West Bengal and the Gahirmatha coast in Orissa are what could be called the stars of mainland India's coastal and marine ecosystems. They are biologically rich and diverse; have a special status under the law for wildlife conservation and are also recognized internationally for their ecological importance. The present rate of sea-level rise if continued, at least 12 islands in the Sunderbans could go completely by 2020. An estimated 70,000 people could be turned into environmental refugees if the problem is not recognized and importantly collective action is not taken to deal with it. Sagar Island, one of the most thickly populated could lose 15 per cent of its land mass in the next decade and the remaining in the next two decades or so. Coastal agriculture will be particularly hit and a 100-cm sea-level rise could cost India more than a thousand million dollars. Little is known of the impact and the changes taking place along other parts of India's coast or even the island groups of Lakshadweep and the Andaman and Nicobar which would be even more susceptible and vulnerable. This is the time to apply the well accepted precautionary principle and initiate urgent action. It is also feared that one fifth of Bangladesh could go under water in the years to come and the intensity and frequency of tropical storms in the Bay of Bengal will increase in the post monsoon period as will the flooding in low lying areas.

Similarly, forest officers in Orissa have noticed and pointed out large scale erosion of the nesting beaches of the Olive Ridley turtles along the famous Gahirmatha coast. While any coastal system is extremely dynamic and the process of erosion and beach formation is a continuous one, the scale of the erosion is reported to have been unprecedented in the last 18 odd months. In these places the sea has moved in by nearly half a kilometer in this short span of time. The causes could be many and while it would be difficult to pinpoint the real reason. There could/should be no denying that global warming would only make the process faster. If this continues unabated we could be watching the decimation of one of the most spectacular natural events experienced on the surface of the earth - the mass nesting of the turtles in these remote beaches.

There was growing evidence that global warming has put the climate in a different mode and therefore the future could be different from the past. Any major changes in rainfall patterns will have far reaching consequences for agriculture and water availability.

Abnormal Rainfall Patterns Forecast in India

Global Warming has been widely predicted to increase extreme weather events including bouts of excessive rain. Extreme rain events have grown both in number and intensity over large parts of Central India in the space of half a century. It has been found that the increase in extreme rainfall events occurred during a period when the waters of the tropical Indian Ocean became warmer producing more water vapour. In addition temperatures over India too increased by 0.5 degrees C in the last 50 years. The capacity of the atmosphere to hold moisture would thereby increase and conditions could become conducive for heavy downpours.

An analysis has been made of daily rainfall patterns during the south-west monsoon from 1951 to 2000 in an area of about 1.4 million square km in Central India extending from Uttar Pradesh in the north to Andhra Pradesh in the south and from Madhya Pradesh in the west to Bihar in the east. Episodes of heavy rainfall where places in Central India received 10 cm or more of rain in a single day during the monsoon had increased at the rate of 10 per cent every decade since the early 1950s. Instances of very heavy rainfall where places received 15 cm or more of rain in a day had more than doubled between 1951 and 2000. The increasing trend in extreme rainfall events over Central India was quite robust and could continue for the next decade or two. As such exceptionally heavy rain could lead to floods and mudslides and the potential for natural disasters would also grow. Further, there could be an increase in severe rainfall activities across an extensive area covering the Western Ghats, Maharashtra, southern Gujarat and parts of Central India by the end of the 21st century.

Threat to Alpine Region in Europe from Global Warming

Recent warming in Europe's alpine region was roughly three times the global average. In Europe, 1994, 2000, 2002 and 2003 were the warmest years on record in the Alps over the past 500 years. Global warming could devastate European ski resorts within decades, forcing lower-altitude resorts to close and threatening winter sports which now attract up to 80 million tourists a year. Germany was most at risk, with the low lying ski areas in Bavaria threatened. In Austria, winter tourism accounting for 4.5 per cent of the national economy is also threatened. In France, the ski industry with a turnover of about 20 billion euros per year may also be threatened. Predictions showed even greater changes in the coming decades, with less snow at low altitudes and receding glaciers and melting permafrost higher up. A two degree rise in temperature would reduce the number of viable slopes that could occur by 2050.

Polar Ice Shelf Breaks Free

A giant ice shelf has snapped free from an island south of the North Pole. Climate change has been cited as a major reason for the event. The 66 sq km long Ayles Ice Shelf broke clear from the coast of Ellesmere Island, about 800 km south of the North Pole in the Canadian Arctic. Within one hour of breaking free, the shelf had formed as a new ice island leaving a trail of icy boulders floating in its wake. It shows that remarkable features of the Canadian North that have been in pace for thousands of years are changing signaling the onset of accelerated change ahead.

Asia and the Challenge of Global Warming

Asia with its high population densities and countries having large populations is particularly at risk from increases in sea level. Almost three quarters of the people currently living in the low elevation coastal zone are Asian. China with an estimated 144 million people, India with more than 63 million and Bangladesh with more than 62 million lead the ranking of countries with the largest populations in the zone. Millions of people around the world including in fast growing Asian cities such as Shanghai and Mumbai are likely to be affected by rising sea levels and stronger storm surges caused by global warming. Over 630 million people accounting for a tenth of the world's population live within 10 meters above the current sea level.

Five billion people and large-scale vegetation in the South Asian region face major threat of global warming. Scientists have warned that 80 per cent of bio-diversity is going to be at an increased risk and people in the region are going to be affected directly or indirectly in the coming years. The general temperature is likely to increase by 2.5°C in the next 50 years and India might face situations like more droughts or floods. North-west India, Pakistan and Nepal are set to see warmer conditions, which will be detrimental for the vegetation and especially the inhabitants in and around the coastal region.

Global warming is the new challenge to rice production and productivity. Asian rice production could decline by an average 4 per cent due to climate change based on the present systems. Altered timing and magnitude of rainfall could induce drought or flood injury to the rice crop. Increased temperatures and/or changes in precipitation could have dramatic impacts on diseases and insects. Enhanced temperatures and enriched carbon dioxide may alter competition between rice and major weeds and the contribution of other organisms to nitrogen fixation in rice fields. Increased atmospheric carbon dioxide may have beneficial impacts on rice growth, the benefits would be lost if the temperatures also rose. That is because increased temperatures shorten the period over which rice grows. Evergreen revolution meant increasing productivity in perpetuity without associated ecological harm, while the green revolution was commodity-centered increase in productivity. The decline in soil health and water quality in rice-based systems is a major global issue. Global warming would have a negative impact on yield and soil fertility.

India and the Challenge of Global Warming

Warming of the climate system is unequivocal due to increased global air and ocean temperatures, widespread melting of snow and ice and rising sea levels. The burning of fossil fuels, some agricultural practices and changes in land use have been generating gases like carbon dioxide, methane and nitrous oxide that trap heat and produce runaway global warming. If the production of these greenhouse gases continued to soar, global temperatures could rise by up to 6.4 °C by the end of this century with far reaching consequences for the climate.

The widespread melting of mountain glaciers and polar ice caps had already contributed to rising sea levels. A sharp rise in sea level could have a considerable impact on India. The United Nations Environment Programme included India among the 27 countries that are most vulnerable to a sea level rise. About a quarter of India's population lives within 50 km of the coastline. The mega cities of Mumbai and Chennai with large and growing populations and huge investments in infrastructure are located on the coast. Besides, much of the costal region has fertile agricultural land. Low level areas such as those in Orissa and West Bengal could be vulnerable to inundation. An increase in sea level could also lead to salt water entering the ground water aquifers on which people depend for drinking water and to irrigate their fields.

Glaciers in the Himalayas feed important rivers such as the Ganga, the Indus and the Brahmaputra that provide water for millions of people as well as for irrigation and industry. The accelerated melting of these glaciers as a result of the earth's warming will have a profound effect on future water availability. The Gangotri glacier, one of the largest in the Himalayas, has been retreating rapidly in recent decades. The smaller glaciers are more sensitive to global warming.

Changes in temperature and rainfall associated with global warming could result in about 80 per cent of the existing forests in the country undergoing a change in the type of vegetation. Such changes were bound to have a very significant impact on the forests and the wildlife they supported. As the Indian forests were already highly fragmented, many species of plants and animals might not be able to cope with climate change and could therefore face extinction.

Human Health

Deaths from all causes are known to rise when temperatures go up, and heart attacks and cardiac arrests are more likely when it is very cold. It was anticipated that global warming would increase deaths during hot temperatures but that this would be compensated for by fewer deaths in the winter. Making air conditioning universally available may reduce heat-related mortality but would, on the other hand, have a perverse effect by enhancing global warming through carbon dioxide emissions from electricity consumption.

Throughout the world, the prevalence of some diseases and other threats to human health depend largely on local climate. Extreme temperatures can lead directly to loss of life, while climate-related disturbances in ecological systems, such as changes in the range of infective parasites, can indirectly impact the incidence of serious infectious diseases. In addition, warm temperatures can increase air and water pollution, which in turn harm human health.

Human health is strongly affected by social, political, economic, environmental and technological factors, including urbanization, affluence, scientific developments, individual behavior and individual vulnerability (e.g., genetic makeup, nutritional status, emotional well-being, age, gender and economic status). The extent and nature of climate change impacts on human health vary by region, by relative vulnerability of population groups, by the extent and duration of exposure to climate change itself and by society's ability to adapt to or cope with the change.

There is a close link between local climate and the occurrence or severity of some diseases and other threats to human health. Extreme temperatures can directly cause the loss of life. Several serious diseases appear only in warm areas. Warm temperatures can increase air and water pollution, which in turn harm human health. Extremely hot temperatures increase the number of people who die on a given day for many reasons: people with heart problems are vulnerable because their cardiovascular system must work harder to keep the body cool during hot weather. Heat exhaustion and some respiratory problems increase. Higher air temperatures also increase the concentration of ozone at ground level. The natural layer of ozone in the upper atmosphere blocks harmful ultraviolet radiation from reaching the earth's surface; but in the lower atmosphere, ozone is a harmful pollutant. Ozone damages lung tissue, and causes particular problems for people with asthma and other lung diseases. Even modest exposure to ozone can cause healthy individuals to experience chest pains, nausea, and pulmonary congestion.

Warm temperatures accelerate the spread of disease primarily by enlarging the geographic range in which disease-carrying animals, insects and microorganisms—as well as the germs and viruses they carry—can survive. With a rise in global temperature, diseases that were previously limited only to tropical areas may show up increasingly in other, previously cooler areas.

Virulence of Mosquito-borne Diseases

Both male and female mosquitoes are nectar feeders, but the females feed on the blood because the normal diet consists of nectar and fruit juice which has no protein and they require the supplemental substances like protein and iron for the development and laying of their eggs. In most female mosquitoes, the mouth parts form a long proboscis for piercing the skin of mammals or in some cases birds or even reptiles and amphibians to suck their blood. Males differ from females, with mouth parts not suitable for blood-sucking.

The female mosquitoes locate their next blood donor victims primarily through scent. They are extremely sensitive to the carbon dioxide in exhaled breath, as well as several substances found in sweat and various body odours. They are believed to be able to track potential prey for tens of meters. Some people attract more mosquitoes than others, apparently based on how they "smell" to a mosquito. Mosquitoes can also detect heat, so they can find warm-blooded mammals and birds very easily once they get close enough. Mosquitoes carry disease-causing viruses and parasites from person to person without catching the disease themselves. When a female mosquito bites, she also injects saliva and anti-coagulants into the blood which may also contain disease-causing viruses or other parasites.

Malaria

Malaria is a vector-borne infectious disease caused by protozoan parasites of the genus Plasmodium. Only four types of the plasmodium parasite can infect humans; the most serious forms of the disease are caused by *Plasmodium falciparum* and *P. vivax*, but other related species *P. ovale* and *P. malariae* can also affect humans. This group of human-pathogenic *Plasmodium* species is usually referred to as malaria parasites. Malaria parasites are transmitted by female *Anopheles* mosquitoes. Malaria is widespread in tropical and subtropical regions, including parts of the Americas, Asia and Africa. Each year, it causes disease in approximately 515 million people and kills between one and three million people. Malaria is commonly associated with poverty, but is also a cause of poverty and a major hindrance to economic development.

Malaria has increased at an alarming rate due to the increase in the global temperature. As the high temperature of the atmosphere spreads from tropical region towards the polar area, the risk of a rise in the spreading of this disease in the regions near the two poles are quite high. Health effects of global warming have a long lasting effect on the living beings on earth.

Global warming will cause or increase incidences of malaria, dengue fever, yellow fever, encephalitis and respiratory diseases throughout the world in coming decades. The news is not good for less developed parts of the world either. Two-thirds of water-borne disease outbreaks such as cholera follow major precipitation events, which are already increasing due to global warming. Malaria, diarrhea, malnutrition and floods cause an estimated 1, 50,000 deaths annually, with Asia accounting for more than half. Malaria-carrying mosquitoes represent the clearest sign that global warming has begun to impact human health, adding they are now found in cooler climates such as South Korea and the highlands of Papua New Guinea. Warmer weather means that mosquito breeding cycles are shortening, allowing them to multiply at a much faster rate, posing an even greater threat of disease. Encephalitis, Malaria and Dengue are the diseases that attack people mainly in warmer temperature. As the temperature increases on earth, it helps the viruses of infectious diseases to grow and get immune to any kind of resistance. In recent times, most of the intensive care units in the American hospitals are filled with patients suffering from the mosquito borne diseases like, Encephalitis, Malaria and Dengue, etc. With the rise in global temperature the diseases spread on a wider area and affect the health of millions.

Dengue Fever

This is also called break bone fever and is a mosquito borne disease. This fever is also related to yellow fever and no vaccine can exactly stop this disease from spreading. Dengue is transmitted by the bite of an *Aedes* mosquito infected with dengue viruses. It occurs in tropical and sub-tropical areas of the world. Dengue viruses occur in most tropical areas of the world. Dengue is common in Africa, Asia, the Pacific,

Australia, and the Americas. It is widespread in the Caribbean basin. Dengue is most common in cities but can be found in rural areas. More than 70% of the disease burden is in South-East Asia and the Western Pacific area. Africa and the Eastern Mediterranean are much less affected. In Latin America and the Caribbean, the incidence and severity of disease are increasing rapidly. Increase in international air travel is facilitating the rapid global movement of dengue viruses. This increases the risk of dengue haemorrhagic fever epidemics by introducing new dengue viruses into susceptible populations. Hemorrhagic dengue fever is a fatal form of dengue and once it spreads like an epidemic it takes many lives. Frost kills adult and larvae of mosquito. Previously this disease was unknown to the polar areas of the earth. Gradually temperature spreads from tropical region towards the poles and presently mosquitoes are spreading this disease to the regions near the north and south poles. Dengue is among the severe health effects of global warming.

All types of dengue virus are re-emerging worldwide and causing larger and more frequent epidemics, especially in cities in the tropics. The emergence of dengue as a major public health problem has been most dramatic in the western hemisphere. Dengue fever has reached epidemic levels in Central America and is threatening the United States. Dengue hemorrhagic fever is also on the rise. Persons who have been infected with one or more forms of dengue virus are at greater risk for the more severe disease. With the increase in all types of virus, the occurrence of dengue hemorrhagic fever becomes more likely.

Chikungunya

Chikungunya virus is an arbovirus of the genus, *Alphavirus* that is transmitted to humans by virus-carrying *Aedes aegypti* mosquitoes. It causes arthritic symptoms and in effect walking becomes difficult. Chikungunya virus is indigenous to tropical Africa and Asia, where it is transmitted to humans by the bite of infected mosquitoes, usually of the genus *Aedes*.

There have been recent outbreaks of chikungunya associated with severe morbidity.

Japanese encephalitis

Japanese encephalitis is a disease caused by the mosquito-borne Japanese encephalitis virus which belongs to the family Flaviviridae. Domestic pigs and wild birds are reservoirs of the virus; transmission to humans may cause severe symptoms. The most important vectors of this disease is the mosquito *Culex tritaeniorhynchus* which feeds on cattle in preference to humans. This disease is most prevalent in South-east Asia and the Far East.

Yellow fever

Yellow fever is a viral disease that is transmitted to humans through the bite of infected mosquitoes. The disease occurs only in sub-Saharan Africa and tropical South America. The "yellow" in the name is explained by the jaundice that affects some patients. The disease is caused by the yellow fever virus, which belongs to the *flavivirus* group. The mosquitoes involved in the transmission of the disease are *Aedes* and *Haemogogus* mosquitoes transmit the yellow fever virus.

The exceptionally high number of cases in Asia of dengue fever, which is also spread by mosquitoes, could be due to rising temperatures and rainfall. Around the world, mosquitoes are moving into elevations previously too cold for them to survive in. As mosquito populations grow due to warmer climate, they can spread vector-borne diseases. In Colombia, mosquitoes that carry dengue fever and yellow fever viruses were previously limited to elevations below 3,300 feet but have recently been found at 7,200 feet. In Mexico, dengue fever has spread above its former limit of 3,300 feet and now can be found at 5,600 feet.

The rise of many infectious diseases and other threats to human health depends in large part on the local climate. Many mosquito-borne diseases are sensitive to climate conditions. Temperature rise is one of the most common factors contributing to the increase of mosquito borne diseases. It is predicted that global warming will increase the risk of infectious diseases, and most particularly those that are prevalent in warm areas of the world. The spreading of disease will be facilitated by mosquitoes and other insects as warmer temperatures will allow them to propagate in areas further north. What this means is that diseases like malaria, dengue fever, yellow fever and encephalitis, which are common in warm weather countries, will begin to invade countries that previously had no reason to be concerned with such health problems. Scientists around the world have been seriously contemplating the effects of global warming on mosquito-borne diseases. Today, scientists and professionals in the medical field are particularly concerned with how climate change will affect the mosquito population and how it will subsequently give potential rise to a variety of dangerous infectious diseases. Global warming is a reality today; hopefully, scientists will find a way to eliminate the problem of mosquito borne diseases before a large scale epidemic takes place.

Bird-flu

Like humans and other species, birds are susceptible to flu. There are many types of bird or avian flu. The most contagious strains, which are usually fatal in birds, are H5 and H7. There are nine different types of H5, some are highly pathogenic, while some are pretty harmless. The type currently causing concern is the "highly pathogenic" Asian strain of the H5N1 virus. Scientists have discovered four different subtypes of H5N1, and there could well be more. However, all are deadly to birds, and can cause disease and death in humans. Humans have been infected but almost all have been poultry workers who have come into intimate contact with birds. H5N1 cannot pass easily from human to human. Migratory wildfowl, notably wild ducks, are natural carriers of the viruses, but are unlikely to actually develop an infection. The risk is that they pass it on to domestic birds, which are much more susceptible to the virus.

Bird-flu was thought only to infect birds until the first human cases were seen in Hong Kong in 1997. Since then it has spread westwards through Asia, the Middle East, Europe and Africa. Humans catch the disease through close contact with live infected birds. Birds excrete the virus in their faeces, which dry and become pulverized, and are then inhaled. Symptoms are similar to other types of flu - fever, sore throats and coughs. People can also develop conjunctivitis. There is no failsafe way of preventing its spread. Proper poultry controls such as preventing wild birds getting in to poultry houses are vital. In addition, monitoring of the migratory patterns of wild birds should provide early alerts of the arrival of infected flocks meaning they could be targeted on arrival. If the virus gained the ability to pass easily between humans the results could be catastrophic. Worldwide, experts predict anything between two million and 50 million deaths. However the mortality rate - which presently stands at around 50% of confirmed cases - could decline as it mutates, they say.

Influenza is not normally a disease that is transmitted by a vector such as a mosquito. But because the virus is relatively environmentally stable, mechanical transmission by insects, such as flies, is a possibility. Since the virus can survive for long periods in the wild, flies might pick it up when they land on feces or infected birds, and then carry it to other animals. Mosquitoes typically transmit diseases, such as West Nile Virus, by ingesting blood from infected animals, then transmitting it to the next animal they feed on.

Bird-flu is likely to spread more quickly as the Earth warms up, but for a different reason. The global warming in concert with excessive development—is contributing to an increased loss of wetlands around the world. This trend is already forcing disease-carrying migrating birds, who ordinarily seek out wetlands as stopping points, to instead land on animal farms where they mingle with domestic poultry, risking the spread of the disease via animal-to-human and human-to-human contact.

Rise in Water Temperature and Emergence of Heat Loving Amoeba

Naegleria fowleri, known as the brain eating amoeba is a free living amoeba lives almost everywhere in lakes, hot springs, even dirty swimming pools grazing off algae and bacteria in the sediment. But, it is typically found warm fresh water, from 25-35 degrees Celsius in an amoeboid or temporary flagellate stage. It belongs to a group called the Percolozoa or Heterolobosea. It can invade and attack the human nervous system; although this occurs rarely, such an infection will nearly always result in the death of the victim.

People become infected when they wade through water and stir up the bottom. If someone allows water to shoot up the nose, the amoeba can latch onto the olfactory nerve. The amoeba destroys tissue as it makes its way up into the brain, where it continues the damage basically feeding on the brain cells. People who are infected tend to complain of a stiff neck headaches and fevers. In the later stages they will show signs of brain damage. Once infected most people have little chance of survival. Some drugs have stopped the amoeba in lab experiments but people who have been attacked rarely survive. This is a heat loving amoeba. As water temperatures go up it does better. In future decades, as temperatures rise we would expect to see more cases.

African Dust Clouds could Bear Infections

Rising temperatures drive the giant clouds of dust whipped up by desert storms in Africa can carry infectious organisms to other continents. Despite being blown more than 4.8 km high and exposed to radiation from the sun, strains of bacteria and fungi survived and were able to grow when they returned to the earth.

Among 40 tests of air samples taken in the mid Atlantic, 24 revealed living microbes including 26 colonies of bacteria and 83 fungi. They included strains capable of causing disease in humans, animals and plants even if the UV kills 90 per cent of the microbes, dust higher up in the cloud will shade those lower down resulting in phenomenal number surviving. The microbes included *Gordona terra* which can cause skin disease in humans, *Massaria rosatii*, which infects sycamore trees, and *Alternaria dauci* causing carrot blight. The increasing temperature resulting from global warming is bound to amplify the problems of diseases to plants and animals.

The global warming is also bad to allergy sufferers. A higher level of carbon dioxide in the atmosphere acts like a super charger for ragweed, the bane of many people with allergies. In greenhouses, when an environment with double the normal carbon dioxide levels, the ragweed plan'ts thrived growing taller and bushier and producing more seeds. The plants also produced 61 per cent more pollen than ragweed plants grown in greenhouses with normal atmosphere. In effect, the implications go beyond just the misery associated with allergies and aggravate asthma.

Increase in Insect Numbers and Other Animals due to Warming

An important consequence of global warming is the rise in insect populations in response to warmer temperatures. This is welcome in case of butterflies. But, it is not so with mosquitoes or any other insects which act pests or the like. It is an established fact that insects are highly adaptable organisms and through evolution have adjusted to a variety of temperatures and environments. Insect species that adapt to warmer climates also will increase their maximum rates of population growth and are likely to have widespread affects on agriculture, public health and conservation.

One of the key mechanisms of evolution, natural selection, causes organisms to evolve in response to a changing environment. Imagine a population with several different variants in it: some individuals happen to be better able to survive and reproduce at higher temperatures than other individuals. Clearly, if the temperature increases, those heat tolerant individuals will have an advantage and will leave more offsprings — and those offsprings will also carry the genes for heat tolerance. Over many generations, this process produces a population with adaptations well-suited for the hotter environment. So long as the population has different genetic variants in it, some better able to survive and reproduce in particular situations than others, the population has the capacity to evolve when faced with a changing environment.

In many cases, these changes are simply non-evolutionary examples of phenotypic plasticity, where an organism expresses different traits depending on environmental conditions. For example, many organisms respond to warmer weather by reproducing earlier and taking advantage of an earlier spring, but this early reproduction is not caused by genetic changes in the population and so is not an example of evolutionary change. Similarly, many species have shifted their ranges in response to this tiny temperature difference, spreading towards the poles, as those habitats warm, but this change in range cannot be traced to a genetic shift in the population and so is not an example of evolution. And still other species simply seem to be on the path to endangerment or extinction as their habitats are degraded and their population sizes drop. However, in a few cases, the species have actually evolved, experienced a change in gene frequency in the population in response to global warming. Interestingly, in those cases, the species are not necessarily becoming more heat tolerant, but are adapting to changes in seasonal timing.

Global warming could pose a greater risk to tropical insects and other species sensitive to the slightest shifts in temperature than to creatures living in the world's tundra. While cold weather animals are used to huge temperature changes, tropical species live under a much smaller temperature range and face a bigger risk of extinction with an increase of just two or four degrees Celsius. In the tropics many species appear to be living at or near their thermal optimum, a temperature that lets them thrive. But once temperature gets above the thermal optimum, fitness levels most likely decline quickly and there may not be much they can do about it.

Polar bears can develop thicker fur to shield them from

freezing temperatures; tropical species must use other tactics to protect themselves from higher temperatures such as staying out of direct sunlight or burrowing into the soil. But hiding from the sun could prove useless to tropical animals already living so close to their temperature comfort zone as the warmer weather could come too fast for their physiologies to adapt. Many tropical species can only tolerate a narrow range of temperatures because the climate they experience is pretty constant throughout the year. Unfortunately, the tropics also hold the large majority of species on the planet.

Canadian squirrels are evolving to take advantage of an earlier spring and are breeding sooner, which allows them to hoard more pinecones for winter survival and next year's reproduction. Squirrels with genes for earlier breeding are more successful than squirrels with genes for later breeding. European great tits are also evolving different breeding times. Birds that are able to adjust egg-laying to earlier in the spring can time hatching so that it coincides with greater food (caterpillar) abundance and with recent climatic changes, the caterpillars have been maturing earlier in the spring. Birds with genes for more flexible egg-laying times are more successful than birds with less flexibility in their egg-laying. Another European bird, the blackcap, has been evolving due to changes in its migration patterns. One North American mosquito species has evolved to take advantage of longer summers to gather resources while the weather is good. Mosquitoes with genes that allow them to wait longer before going dormant for the winter are more successful than mosquitoes that go dormant earlier. In a sense, these populations are the lucky ones. Small animals tend to have large population sizes and short generation times and that bodes well for their ability to evolve along with a changing environment. Large population size means that the species are more likely to have the genetic variation necessary for evolution, and having a short generation time means that their rate of evolutionary change may be able to keep pace with environmental change. However, other species may not be so lucky: larger animals tend to have longer generation times and so evolve more slowly and larger animals also tend to have smaller population sizes, which means that their populations are simply less likely to contain the gene versions that would allow the population to adapt to warmer climates. If global warming continues, such species may come face to face with extinction, as the environments to which they have been adapted over the course of thousands or millions of years change right out from underneath them in the course of a few decades.

Phytoplankton Respond by Scaling Down Productivity

Microscopic plankton comes in animal and plant forms. The plants are known as phytoplankton. They lie at the base of the marine food chain because they convert sunlight and carbon dioxide into organic carbon - food for everything else. Smaller animals such as shrimp-like krill feed on plankton and are themselves eaten by larger organisms, from small fish to the biggest whales. Without phytoplankton, the oceans would soon be marine deserts. Phytoplanktons are also important because of the role they play in the carbon cycle, which determines how much carbon dioxide - the most important greenhouse gas - ends up in the atmosphere to cause global warming. Huge amounts of carbon dioxide from the atmosphere, which dissolves in the oceans, are absorbed by phytoplankton and converted to organic carbon. When the phytoplanktons die their shells and bodies sink to the seabed, carrying this carbon with them. Phytoplankton therefore acts as a carbon "sink" which takes carbon dioxide from the atmosphere and deposits the carbon in long-term stores that can remain undisturbed for thousands of years. If the growth of phytoplankton is interrupted by global warming, this ability to act as a buffer against global warming is also affected - leading to a much-feared positive feedback.

The ubiquitous phytoplankton, plays a significant and previously unknown role in warming the planet by capturing and absorbing the Sun's radiation. Scientists show that solar radiation that would otherwise be reflected back into space was being absorbed by phytoplankton, contributing to global warming. Some types of phytoplankton species absorb more radiation than others, and there are some types that reflect the Sun's radiation rather than absorb it. Changes in phytoplankton concentrations can affect the Earth's surface reflection of radiation from the Sun, which in turn influences how the planet's climate system interacts with human-generated greenhouse gases and aerosols. There is a need to consider the point of 'iron fertilization' of the oceans if ever used to reduce global warming. If the oceans were seeded with vast quantities of iron sulphate (the lack of which limits phytoplankton growth), it would trigger a boom in phytoplankton rates and allow the organisms to draw more carbon dioxide from the atmosphere. Iron fertilization would probably increase populations of phytoplankton. Instead of reducing global warming, it would contribute to warming the ocean by absorbing more radiation. It would be important to calculate how much the initial decrease in temperature from fertilization would be offset by the increase from radiation. The vital plankton of the oceans can be starved of nutrients as a result of the seas getting warmer to handle the effect of global warming. They believe the findings have catastrophic implications for the entire marine habitat, which ultimately relies on plankton at the base of the food chain.

Global warming, which is causing the temperature of the sea surface to rise, will also interfere with the vital upward movement of nutrients from the deep sea. These nutrients, containing nitrogen, phosphorus and iron, are vital food for phytoplankton. If the supply is interrupted with the plants die off, it prevents them from absorbing carbon dioxide from the atmosphere. Global warming of the surface layers of the oceans reduces the upward transport of nutrients into the surface layers. This generates chaos among the plankton.

Global warming is creating an ocean famine in swaths of tropical and sub-tropical seas. If warming continues, fish stocks in tropical and sub-tropical regions will drop significantly. In some ocean regions, phytoplanktons respond to rising temperatures by scaling down their productivity by 30 per cent or more. With less production at the bottom of the food chain, fish and other large ocean creatures have less to eat. The changes occur because warmer surface temperatures lead to changes in the flow of ocean currents that deliver nutrient-rich water from the cooler depths to the surface. About half of the production generated by the world's living organisms is done by phytoplankton, microscopic green plants which operate in the top 100 to 200 meters of the ocean where light levels are high enough for photosynthesis. Each day they pull in more than 100 million tons of CO_2. Almost the whole food chain of the open ocean depends on these plants.

Climate change has profound implications for virtually all aspects of human well being, from jobs and health to food security and peace within and among nations. It could exacerbate tensions and trigger conflicts across the world by worsening food, water and land resource shortage and increasing the number of environmental refugees.

Collapse of Ecosystems

Humans are living well beyond their ecological means and are now exhausting natural resources at an unprecedented rate. The planet's capacity to provide food, fiber and timber and absorb carbon dioxide is diminishing; the use of resources is 25 per cent more than the ability of nature to renew them in a year. This ecological overshoot which has been growing steadily for nearly 40 years will on present trends be 100 per cent by 2050 making the likelihood of large-scale ecosystem collapse likely, and conflict and political tension certain. Effectively the earth's regenerative capacity can no longer keep up with demand. People are turning resources into waste faster than nature can turn waste back into resources. Humanity is no longer living at nature's interest but is drawing down its capital. This growing pressure on ecosystems is causing habitat destruction or degradation and is threatening both biodiversity and human well being.

The geographic ranges of most plant and animal species

are limited by climatic factors, including temperature, precipitation, soil moisture, humidity, and wind. Any shift in the magnitude or variability of these factors in a given location will impact the organisms living there. Species sensitive to temperature may respond to a warmer climate by moving to cooler locations at higher latitudes or elevations. Although the response to warming is generally understood, it is difficult to predict how concurrent changes in other climatic factors also affect species distributions. Despite the uncertainties, ecological models predict that the distribution of world biomes will shift as a result of the climate changes associated with increased greenhouse gases. The distribution and size of the populations of plants and animals within those biomes will also change, with potential consequences for the functioning of ecosystems and for humans who are dependent on many ecosystem goods and services.

Global biome models have been used to evaluate the changes in vegetation distribution likely to occur with a changing climate. The models generaliy predict a pole ward shift of the northern hemisphere taiga, boreal conifer, and temperate mixed forests belts. One limitation of the global models is that the output represents a vegetation distribution that is in equilibrium with climate, a condition that is unlikely to occur in the next century. The spread of tree species involves several factors, including dispersal, regeneration on a suitable site, maturation, and seed production. If climate changes faster than trees can disperse to new, more suitable areas, the composition of the forest may change and the survival of some species could be at risk. Global-scale models are also inadequate to evaluate the indirect effects of climate, such as disturbances from pests, disease, fire, flooding, and wind damage.

Many animals are able to respond to climate at a faster rate than plants. For those animals that do not migrate, a distribution change in response to a warming trend would occur at the population level as a result of changes in the ratios of extinctions to colonization at the northern and southern boundaries of the range. A northward range shift would thus be reflected in either a net extinction at the southern boundary or a net colonization at the northern boundary. Range shifts in areas with regional warming trends have been reported in alpine plants, butterflies, birds, marine invertebrates, and mosquitoes.

Factors other than climate may limit the extent to which organisms can shift their ranges. Physical barriers such as mountain ranges or extensive human settlement may prevent some species from shifting to more suitable habitat. In the case of isolated mountain top species, there may be no new habitat at higher elevation to colonize. Even in cases where no barriers are present, other limiting factors such as nutrient or food availability, soil type, and the presence of adequate breeding sites may prevent a range shift. Although tree line will probably increase in elevation as climate warms, soils at higher elevations are often thin and of poor quality and could be inadequate to sustain species from lower elevation sites. In coastal areas, the loss of wetlands and beaches due to sea level rise could destroy sites used by turtles, birds, and marine mammals for breeding and raising young.

Global Warming and Ozone Depletion

Continued global warming will accelerate ozone destruction and increase stratospheric ozone depletion. Ozone depletion gets worse when the stratosphere (where the ozone layer is), becomes colder. Because global warming traps heat in the troposphere, less heat reaches the stratosphere which will make it colder. Greenhouse gases act like a blanket for the troposphere and make the stratosphere colder. In other words, global warming can make ozone depletion much worse right when it is supposed to begin its recovery during the next century. UV-B radiation causes skin cancer, cataracts and immune suppression in both animals and humans. UV-B also damages plants including hardwood forests, and phytoplankton.

Human susceptibility to infections can be further compounded by malnutrition. UV-B radiation from ozone depletion damages both plants and animals. UV-B harms amphibian eggs, midge larvae and trout. Crops that are damaged will reduce food availability. UV-B also can damage mammalian immune systems which makes humans and other animals more susceptible to infectious diseases.

Ozone depletion and global warming have harmful effects on plants and animals. If allowed to continue, our food chain will be seriously disrupted. In Antarctica, there has been upwards of 50 per cent ozone depletion. This means that an unusually high amount of UV-B radiation has reached the Earth's surface in the Antarctic region. UV-B harms the productivity of phytoplankton, thereby reducing the available food for animals that feed on phytoplankton. Krill eat phytoplankton and penguins eat krill. From a climate change perspective, phytoplankton normally absorbs a lot of carbon from the air. As phytoplankton dies from UV-B radiation, this carbon is no longer absorbed. This means that more carbon will be left in the atmosphere, contributing to more global warming. More global warming can increase ozone depletion, which kills more phytoplankton, and the process repeats itself.

Phenology

It is the study of the times of recurring natural phenomena. It is principally concerned with the dates of first occurrence of natural events in their annual cycle. Examples include the date of emergence of leaves and flowers, the first flight of butterflies and the first appearance of migratory birds, the date of leaf colouring and fall in deciduous trees, the dates of egg-laying of birds and amphibia, or the timing of the developmental cycles of temperate-zone honey bee colonies. Phenological records help alert us about the events of nature and provide interesting comparisons between years and among different geographic regions. If recorded carefully and consistently, these records also have scientific value for understanding the interactions between organisms and their environment and for assessing the impacts of climate change.

Shifting plant phenology in recent decades establishes that species and ecosystems are already responding to global environmental change. Earlier flowering and an extended period of active plant growth across much of the northern hemisphere have been interpreted as responses to warming. However, several kinds of environmental change have the potential to influence the phenology of flowering and primary production.

That climate changes are indeed occurring, and are in large part due to human activities, is by now widely accepted. Climate variability and extreme weather events are also expected to increase, affecting not just people but also the intricate timing of development phases of all living organisms that are part of the Earth's natural systems. This latter point is crucial: if our biological and climate calendars start to run out of synchronization due to these global changes, there will be significant ecological and socio-economic consequences that will seriously affect humans. For example, it has been well documented that, during the 20th century, many parts of the world experienced warmer surface air temperatures that caused earlier and longer growing seasons. But what if this means that flowers start blooming before any insects have hatched or any migrating pollinators are around, to pollinate them? It is easy to see that ultimately, such changes could filter up the food chain to have drastic effects on our food supply. Phenological data can help us document and measure the occurrence of such changes.

Phenological information is useful for more than just understanding current conditions of climate change. When spatially extensive sets of phenological data are integrated with short- and long-term climatic forecast models, powerful tools emerge to assist human adaptation to ongoing and future climate change. Given sufficient observations and understanding, these tools can be used as a predictor for other processes and variables of importance at local to global scales. Ultimately, phenological data could drive a variety of ecological forecast models with both theoretical and practical applications, enabling answers to critical questions about our changing planet and how humans can best adapt to this change through increased understanding of earth systems interactions on a scale never before possible.

Airborne-pollen

Pollen, another air contaminant, is likely to increase as temperatures and carbon dioxide levels rise. Warmer temperatures will help certain plants, like ragweed, produce additional pollen, as the growing season lengthens and day and night-time temperatures increase. Ozone, a major component of smog, is formed when tailpipe and smokestack emissions cook in the hot summer sun, so increased summer temperatures will also increase smog and ozone pollution. Both pollen and smog contribute to asthma and allergies, and ozone pollution has additional impact on lung and heart health. With increasing temperatures, allergy seasons may turn out to be allergy years.

Each year, as the world's flora goes sex-mad, airborne pollen—the flowering plant's equivalent of semen—along with spores from fungi and non-flowering plants unleash an epidemic of snuffles, sneezes, itchy eyes, or worse. Landing in the lush mucous membrane of a human nose, pollen grains behave as if they had found a moist female flower part, and unleash their surface proteins. In susceptible individuals these alien molecules set off an immune cascade centered on antibodies belonging to the class known as immunoglobulin E. The main aim in life of these antibodies is to help mast cells and eosinophils take action against parasites that penetrate the gut wall. The eosinophils release poisons to kill the parasites, and the mast cells squirt out histamine to trigger a flood of mucus to wash them away. When IgE mistakes pollen for a dangerous parasite, it leads to hay fever or "allergic rhinitis"—essentially a snuffly nose. But not all pollen is to blame. Plants that spread their sex cells on the wind, rather than relying on insects to do the job, trigger most allergies. Allergenic plants also have to be common, so that the immune systems of susceptible individuals can be sensitized through repeated exposure.

The immune system reacts to pollutants in the air by

triggering a chemical reaction that causes inflammation and irritation in many people, especially those sensitive to such particles. In stagnant air, pollutant particles linger and can cause problems for people. Pollen, however, sits still. But good air movement clears away air pollutants and stirs up pollen that causes more problems for people afflicted with allergies and asthma. Pollen triggers a different immunological response in the body than pollutants do by activating certain antibodies that start histamine production. Histamine causes the inflammation in this case.

Urban landscapes with litter-free horticulture are heavily loaded with wind-pollinated dioecious male cultivars (clones) of trees and shrubs in some parts of the world. These modern landscape trees result in surrounding air with unnaturally large amounts of allergenic pollen. Because the "messy" urban female trees are now so rare, almost none of this pollen is now trapped, removed from the air and turned into seed. Under normal carbon dioxide levels these male cloned trees will always produce abundant amounts of pollen. Under increased levels of carbon dioxide, they produce considerably more. The increase in temperature itself also results in increased pollen production. Under stress conditions, male plants are able to take up more water than are females. Under stress conditions, such as drought, male trees are also able to hold onto the water they already have better than are female plants. Where there are abundant water and soil nutrient sources, the increases in carbon dioxide levels in the air will result in larger urban trees, which if they are allergy trees, will be capable of producing ever more pollen. Increases in carbon dioxide increase plant growth but only if there is enough available extra water and nitrogen in the soil to support this additional growth. When the supplies of water and nutrients are not adequate to support this added carbon dioxide-induced growth, interesting physiological things happen in plants. Foremost, it is an added stress on the plants and stress often results in an increase in unusual reproduction factors.

In avenue plantations in urban areas, the trees chosen

are usually copious pollen producers and also their pollen is allergic. With increased carbon dioxide levels, these trees produce much more pollen which upon release into the atmosphere changes the air quality. Added to this, pollutants which are always there in the atmosphere will interact with pollen grains and amplify the problems of allergy and asthma in people.

Energy Crisis

An energy crisis is any great bottleneck in the supply of energy resources to an economy. It usually refers to the shortage of oil and additionally to electricity or other natural resources. An energy crisis may be referred to as an oil crisis, petroleum crisis, energy shortage, electricity shortage or electricity crisis. There are currently 98 oil producing countries in the world, of which 64 are thought to have passed their geologically imposed production peak, and of those 60 are in terminal production decline. The per capita yearly consumption of oil is 25 barrels for America, 10-12 barrels for Japan and Italy, 1.5 barrels for China and 0.8 barrels for India. The global average is less than 4.7 barrels per capita yearly. China, and to a lesser extent India, are following the US in rapidly expanding their economies using the technologies of cheap oil. The high prices for oil today are creating a widespread awareness that there is not enough oil being pumped today to supply everybody. Poor Third World countries are already priced out of the market.

The United States, with less than 5 per cent of the world's population, has long been the largest energy consumer in the world. Americans account for one-quarter of the global carbon dioxide emissions today, including nearly half of the increase in these emissions since 1990. These emissions are unmistakably beginning to tie our excessive energy use to the long-term health of the planet. Even more worrisome is the explosion of consumption worldwide. Globalization is tempting the rest of the world to mimic the American lifestyle, complete with private automobiles, large houses and disposable products.

Energy is used not only to run appliances and cars, but also to make virtually everything we buy – from light bulbs to plastic bottles to computers. If just half the global population emulated our level of energy consumption, the consequences would be devastating.

Humans, with a few notable exceptions, have not yet demonstrated willingness and desire to make any sacrifice to assure a successful transition to renewable energy sources before fossil fuels become extremely scarce and expensive. For any country, to protect their populations and future generations against energy shortages and higher energy prices, then it is imperative to reduce dependence on foreign sources of fuel. Fuel shortages may occur at any time due to hostile actions by foreign nations and they will occur because growing populations are competing for the rapidly dwindling remainder of fossil fuels. Reducing the potentially catastrophic accumulation of greenhouse gasses would be a side benefit of using fossil fuels much more efficiently. Reducing fuel consumption will also reduce emissions of toxic mercury, lead, arsenic, nitrous oxides, and other pollutants that are released into the environment during combustion of fossil fuels.

The current energy crisis has a strong connection to global warming. Both problems are rooted in the unsustainable use of fossil fuels as our primary energy source. Any safe, sustainable and long-term solution to our energy crunch is guaranteed to be the right solution for global warming as well. Conversely, prolonging the dependence on fossil fuels by further oil and gas production without seriously shifting the focus to conservation, using energy more efficiently and developing more renewable energy sources will not only propel closer to potentially disastrous global warming scenarios, but could bring on future energy crises just as serious as this one.

3

CLIMATE CHANGE

The Earth's climate has changed throughout history. From glacial periods or ice ages where ice covered significant portions of the Earth to interglacial periods where ice retreated to the poles or melted entirely - the climate has continuously changed. Climate change is the most severe problem that the present humanity is facing today and more serious even than the threat of terrorism.

Changes in the shape of the Earth's orbit as well as the Earth's tilt and precession affect the amount of sunlight received on the Earth's surface. These orbital processes which function in cycles of 1, 00,000 (eccentricity), 41,000 (tilt), and 19,000 to 23,000 (precession) years – are thought to be the most significant drivers of ice ages. Changes occurring within or inside the sun can affect the intensity of the sunlight that reaches the Earth's surface. The intensity of the sunlight can cause either warming for stronger solar intensity or cooling for weaker solar intensity.

Volcanoes can affect the climate because they can emit aerosols and carbon dioxide into the atmosphere. Volcanic aerosols tend to block sunlight and contribute to short-term cooling. Aerosols do not produce long-term change because they leave the atmosphere not long after they are emitted. The eruption of the Tambora Volcano in Indonesia in 1815 lowered global temperatures and historical accounts in New England describe 1816 as "the year without a summer."

The heating or cooling of the Earth's surface can cause changes in greenhouse gas concentrations. For example, when global temperatures become warmer, carbon dioxide is released from the oceans. When changes in the Earth's orbit trigger a warm period, increasing concentrations of carbon dioxide may amplify the warming by enhancing the greenhouse effect. When temperatures become cooler, carbon dioxide enters the ocean and contributes to additional cooling. That is, during warm interglacial periods, carbon dioxide levels have been high and during cool glacial periods, carbon dioxide levels have been low.

The heating or cooling of the Earth's surface can cause changes in ocean currents. Because ocean currents play a significant role in distributing heat around the Earth, changes in these currents can bring about significant changes in climate from region to region.

Interglacial climates (such as the present) tend to be more stable than cooler, glacial climates. For example, the climate during the current and previous inter-glacials has been more stable than the most recent glacial period. This glacial period was characterized by a long string of widespread, large and abrupt climate changes. Abrupt or rapid climate changes tend to frequently accompany transitions between glacial and interglacial periods and vice versa. While abrupt climate changes have occurred throughout the Earth's history, human civilization arose during a period of relative climate stability. During the last 2,000 years, the climate has been relatively stable. Scientists have identified three departures from this stability, known as the Medieval Climate Anomaly, the Little Ice Age and the Industrial Era. The Medieval Climate Anomaly refers to relative warmth that existed between roughly 900 and 1300 AD in Europe, Greenland and Asia experienced. Historical accounts and other evidence document the warmth that occurred in some regions, the geographical extent, magnitude and timing of the warmth during this period is uncertain. The American West experienced very dry conditions around this time. The global existence of a "Little Ice Age" which was not a true "ice age" since major ice sheets did not develop occurred between about 1500 and 1850. The Industrial Era refers to the emergence of an additional warm period in the last 100 years, coinciding with substantially increasing emissions of greenhouse gases from human activities.

In addition to changes in the atmosphere's composition, changes in the land surface can have important effects on climate. A change in land use and cover can affect temperature by changing how much solar radiation the land reflects and absorbs. Processes such as deforestation, reforestation, desertification and urbanization often contribute to changes in climate (including temperature, wind and precipitation) in the places they occur. These effects may be significant regionally, but reduced when averaged over the entire globe. Changes in land cover and land use can also affect the amount of carbon dioxide taken up or released by the land surface.

Greenhouse gas concentrations in the atmosphere will increase during the next century unless greenhouse gas emissions decrease substantially from present levels. Increased greenhouse gas concentrations are very likely to raise the Earth's average temperature, influence precipitation and some storm patterns as well as raise sea levels. The magnitude of these changes, however, is uncertain. The amount and speed of future climate change will ultimately depend on:

- Whether greenhouse gases and aerosol concentrations increase, stay the same or decrease.
- How strongly features of the climate such as temperature, precipitation and sea level respond to changes in greenhouse gas and aerosol concentrations.
- How much the climate varies as a result of natural influences from volcanic activity and changes in the sun's intensity and its internal variability.

Climate-Sensitive Diseases

The process of climate change, including both increases

in global average temperatures and changes in other climate characteristics such as the spatial and temporal distribution of precipitation, has important implications for human health. The warming of the planet will be gradual, but the effects of extreme weather events – more storms, floods, droughts and heat waves – will be abrupt and acutely felt. Both trends can affect some of the most fundamental determinants of health: air, water, food, shelter and freedom from disease. Climate change is a global phenomenon, but its consequences will not be evenly distributed. In short, climate change can affect problems that are already huge, largely concentrated in the developing world, and difficult to control.

There is a close link between local climate and the occurrence or severity of some diseases and other threats to human health. Extreme temperatures can directly cause the loss of life. Moreover, several serious diseases appear only in warm areas. Finally, warm temperatures can increase air and water pollution, which in turn harm human health. The most direct effect of climate change would be the impacts of hotter temperatures themselves. Extremely hot temperatures increase the number of people who die on a given day for many reasons: People with heart problems are vulnerable because their cardiovascular system must work harder to keep the body cool during hot weather. Heat exhaustion and some respiratory problems increase. Higher air temperatures also increase the concentration of ozone at ground level. The natural layer of ozone in the upper atmosphere blocks harmful ultraviolet radiation from reaching the earth's surface; but in the lower atmosphere, ozone is a harmful pollutant. Ozone damages lung tissue, and causes particular problems for people with asthma and other lung diseases. Even modest exposure to ozone can cause healthy individuals to experience chest pains, nausea, and pulmonary congestion.

Climate change endangers human health. Climate-sensitive impacts on human health are occurring today. They are attacking the pillars of public health. And they are providing a glimpse of the challenges public health will have to confront on a large scale. Human beings are already exposed to the effects of climate-sensitive diseases and these diseases today kill millions. They include malnutrition, which causes over 3.5 million deaths per year, diarrhoeal diseases, which kill over 1.8 million, and malaria, which kills almost 1 million.

Warmer temperatures in the East African highlands, for instance, have created more favorable conditions for mosquito populations, thus increasing malaria. Thousands died in unexpected heat waves in Europe during the summer of 2003. Given that climate-sensitive diseases, such as malnutrition and malaria, are concentrated in the developing world, thousands in Asia are therefore at greater risk of acquiring these diseases that have already killed millions. Small island states are likely the countries most vulnerable to climate variability and long-term climate change. They will experience warmer temperatures and changes in rainfall, soil moisture budgets, prevailing winds (speed and direction), and patterns of wave action. Small island states share many features that constrain their ability to adapt to current climate variability and future climate change, including their small or very small physical size, remoteness from major land masses, limited natural resources (often with unique animal and plant life), vulnerability to natural disasters and extreme weather events, economies sensitive to external shocks, populations with high growth rates and densities, poorly developed infrastructure, and limited financial and human resources. Some islands have significant emigration and "brain drain." Small island states also display wide diversity: the islands differ in geologic type, size, elevation, soil composition, drainage characteristics, and natural resources. Some of the larger islands have significant elevation, whereas others are low-lying small coral atolls with limited or no land for evacuation in times of natural disaster. Natural resources range from scarce to abundant. Some islands have abundant surface water, whereas others are completely dependent on groundwater; water requirements are just as diverse. Social, cultural, and economic settings also vary. Some islands have large commercial or industrial centers, and others have extensive agriculture. Infrastructure, including health infrastructure, is sometimes poorly developed. Human communities range from large densely populated cities to small villages and dispersed rural populations. Many small island states are particularly vulnerable to tropical cyclones, storm surges, flooding, and drought. These events can have both short-and long-term effects on human health, including drowning, injuries, increased disease transmission, decreases in agricultural productivity, and an increased incidence of common mental disorders. Because these potential impacts are complex and far-reaching, the true health burden is rarely appreciated. Droughts and floods also can affect health through deterioration of water quality and quantity. Tropical cyclones, especially during or after a drought, also may predispose islands to wildfires, with their associated health consequences.

The distribution and abundance of many vectors may be affected by even small changes in ambient temperature and precipitation, or by changes in vegetation, host populations, and water availability, especially at the margins of vector distribution. Many small island states lie in tropical or subtropical zones with weather conducive to the transmission of diseases such as malaria, dengue, filariasis, and schistosomiasis. The rates of many of these diseases are increasing in small island states for a number of reasons, including poor public health practices, inadequate infrastructure, poor waste management practices, increasing global travel, and changing climatic conditions. In addition to weather and climate factors, social aspects such as culture and traditions are important in disease prevalence. High-priority diseases of concern for small island states include malaria, dengue, diarrheal disease/typhoid, heat stress, skin diseases, and acute respiratory infections and asthma. Outbreaks of these diseases can be costly in lives and economic impacts. Small island states also face health-related problems due to sea-level rise, including coastal flooding; exacerbated storm surges; damaged coastal infrastructure; salination of island fresh water; damage to coastal ecosystems, coral reefs, and coastal fisheries; and population displacement.

Climate change may increase the risk of some infectious diseases, particularly those diseases that appear in warm areas and are spread by mosquitoes and other insects. These "vector-borne" diseases include malaria, dengue fever, yellow fever, and encephalitis. Also, algal blooms could occur more frequently as temperatures warm – particularly in areas with polluted waters – in which case diseases such as cholera that tend to accompany algal blooms could become more frequent. Higher temperatures, in combination with favorable rainfall patterns, could prolong disease transmission seasons in some locations where certain diseases already exist. In other locations, climate change will decrease transmission via reductions in rainfall or temperatures that are too high for transmission.

Monitoring the potential impacts of climate change on health is important for a number of reasons. The provision of epidemiological data is needed to inform policy-makers about the magnitude of effects. Climate change is currently not seen as a priority within the health sector, although different international mechanisms have been set up to sensitize the public health community. As part of surveillance systems, data can help to determine the requirements for and the effectiveness of preventive and adaptive strategies. The monitoring of climate change impacts requires a more holistic approach to infection etiology, examining the possible influence of climate both on the environmental sources of pathogens and on humans. Monitoring is required to detect early effects, promote better research, assist in building integrated assessment models, improve and evaluate adaptation strategies and inform policy makers and the public. Monitoring involves identification of climate sensitive diseases, migration, life styles, land use and adaptation strategies. The criteria to select the climate sensitive diseases should include the following: strength of evidence/climate sensitivity, potential magnitude of the effects, current availability of data including feasibility and cost of collection and short-term benefit of the monitoring process.

Air Quality

Emissions of chemical constituents into the atmosphere by human activity (air pollution) are nothing new. Air pollution has been known to exist since the ancient age of the nomads. The movement of nomads from place to place, never staying at any given location for any length of time was not accidental, it was due to the foul odours generated by plant, animal and human waste. The concerns, however, since the 1950s have been the amount and variety of air pollutants generated through urbanization and industrialization. While some trace gases such as carbon dioxide, sulfur dioxide and ozone are produced from both natural and human-made sources, others such as chlorofluorocarbon compounds are entirely due to human activity.

Clearly, as with any other field of scientific endeavour, in the last 50 years, research on the effects of air pollutants on agriculture has made major advances. Emphasis has shifted from studying the local scale effects of primary pollutants such as sulfur dioxide, to long-range transport and regional scale effects of secondary pollutants such as ozone. Similarly, traditional views of examining the short-term visible injury on the foliage of sensitive crop species have changed to studying the long-term effects on growth and yield.

Climate change is expected to contribute to some air quality problems. Respiratory disorders may be exacerbated by warming-induced increases in the frequency of smog events and particulate air pollution. Ground-level ozone can damage lung tissue, and is especially harmful for those with asthma and other chronic lung diseases. Sunlight and high temperatures, combined with other pollutants such as nitrogen oxides and volatile organic compounds, can cause ground-level ozone to increase. Climate change may increase the concentration of ground-level ozone, but the magnitude of the effect is uncertain. Another pollutant of concern is "particulate matter," also known as particle pollution. Particulate matter is a complex mixture of extremely small particles and liquid droplets. When breathed in, these particles can reach the deepest regions of the lungs. Exposure to particle pollution is linked to a variety of significant health problems. Climate change may indirectly affect the concentration of particulate matter pollution in the air by affecting natural or "biogenic" sources of particulate matter such as wildfires and dust from dry soils.

Air pollution affects the regional and global climate both directly and indirectly. Ozone in the lower layers of the atmosphere contributes to global warming even more than some greenhouse gases included in the Kyoto Protocol, and particulate matter in the atmosphere also has important climate impacts. However, although black carbon, or soot particles, has a warming effect, other particles, for instance sulphates and nitrates, may cool the climate. The current high levels of sulphates and nitrates mask the effects of climate change to some degree. Through cuts in sulphur and nitrogen emissions necessary to protect human health and the environment the climate impacts of the greenhouse gases may actually show more quickly. On the other hand, measures to cut black carbon emissions, for instance from diesel combustion, will have double benefits, protecting both human health locally and also the climate regionally and worldwide.

Methane has a direct negative impact on climate and it contributes to ground-level ozone levels. Methane emissions have grown very rapidly since pre-industrial times. Cutting these emissions will reduce health- and ecosystem-damaging ozone levels and reduce the extent of climate change. While indications of the climate impacts of increasing greenhouse gas concentrations can already be seen in the rise of mean temperatures and the increase in the numbers of extreme climate events (floods and droughts), most impacts are likely to happen over the next 50-100 years. Some gases, like carbon dioxide, stay in the atmosphere for a very long time, so measures to reduce emissions only start to show an effect after a few decades. In contrast, ozone, black carbon and methane can be controlled to show effects much sooner (10-20 years). Cutting these pollutants could help reduce some climate impacts while waiting for longer-term measures to pay off. Besides such links between atmospheric effects, there is also a strong link between the sources of emissions. Energy production and transport are responsible for most CO_2 emissions and much of the air pollution. Cutting energy consumption and car use will therefore have double benefits. Synergies can also be found in agriculture: cutting ammonia emissions could lead to an increase of some greenhouse gas emissions, but the same reduction levels can also be achieved by an integrated strategy that will even cut some of the greenhouse gases.

Agriculture and Food Supply

Agriculture is highly sensitive to climate variability and weather extremes, such as droughts, floods and severe storms. The forces that shape our climate are also critical to farm productivity. Human activity has already changed atmospheric characteristics such as temperature, rainfall, levels of carbon dioxide and ground level ozone. While food production may benefit from a warmer climate, the increased potential for droughts, floods and heat waves will pose challenges for farmers. Additionally, the enduring changes in climate, water supply and soil moisture could make it less feasible to continue crop production in certain regions. Climate variability and change also modify the risks of fires, pest and pathogen outbreak, negatively affecting food, fiber and forestry. The factors such as average temperature increase, change in rainfall amount and patterns, rising atmospheric concentrations of carbon dioxide, pollution levels such as tropospheric ozone and change in climatic variability and extreme events directly connect climate change and agricultural productivity. An increase in average temperature can lengthen the growing season in regions; adversely affect crops in regions where summer heat already limits production; increase soil evaporation rates and increase the chances of severe droughts.

Changes in rainfall can affect soil erosion rates and soil moisture, both of which are important for crop yields. Precipitation will increase in high latitudes, and decrease in most subtropical land regions—some by as much as about 20 per cent.

Increasing atmospheric carbon dioxide levels, driven by emissions from human activities, can act as a fertilizer and enhance the growth of some crops such as wheat, rice and soybeans. Carbon dioxide can be one of a number of limiting factors that, when increased, can enhance crop growth. Other limiting factors include water and nutrient availability. Carbon dioxide fertilization will have a positive impact on some crops; other aspects of climate change may temper any beneficial carbon dioxide fertilization effect.

Higher levels of ground level ozone limit the growth of crops. Since ozone levels in the lower atmosphere are shaped by both emissions and temperature, climate change will most likely increase ozone concentrations. Such changes may offset any beneficial yield effects that result from elevated carbon dioxide levels.

Over the past fifty years, human ingenuity has led to technological advances in agriculture that have dramatically increased crop yields. However, despite these improvements, agriculture is still highly dependent on climate since solar radiation, temperature, and precipitation are the main drivers of crop growth. Since the industrial revolution, humans have been changing the global climate by emitting high amounts of greenhouse gases into the atmosphere, potentially resulting in higher global temperatures, changed hydrological regimes, and increased climatic variability. It seems obvious that any significant change in climate on a global scale should impact local agriculture and thereby affect the world's food supply. Overall, climate change, including global warming and increased climate variability, could result in a variety of impacts on agriculture. Some of these effects are biophysical, some are ecological, and some are economic. They include:

- A shift in climate and agricultural zones towards the

poles.

- Changes in production patterns due to higher temperatures.
- A boost in agricultural productivity due to increased carbon dioxide in the atmosphere.
- Changing precipitation patterns.
- Increased vulnerability of the landless and the poor.

These harmful outcomes of climate change in developing countries and potentially positive outcomes in developed countries will probably increase the gap in wealth, access to food, and health between rich and poor countries. This will affect the worldwide economy as emerging trade-partners are lost, hunger increases, and refugees leave regions harmed by global warming in search of food and resources.

Higher concentrations of carbon dioxide enable plants to use water more efficiently and grow more, the so called "carbon fertilization effect". Of course this benefits weeds as well as food crops; the detrimental effects of other fossil fuel emissions will offset some of the benefits of high carbon dioxide concentrations.

Production could be affected by changes in the prevalence and distribution of livestock pests, and changes in grazing and pasture productivity. Livestocks are sensitive to stress from warmer, drier conditions, as well as any reduction in the quality and availability of their food and water. Where winters become warmer it may make it easier for range livestock to survive.

A rise in sea levels is likely to reduce the amount of land available for agriculture. Rises in sea level threaten ecosystems such as mangrove forests, which are especially diverse. Slight changes in environmental variables, such as temperature, salinity, wind speed and direction, ocean currents, and predators can sharply alter the abundance of fish populations.

The impact of future climate change on food production is variable and the global yields are expected to decrease. The extent to which agricultural systems are affected might depend on the amount of warming that takes place. In the mid latitudes, if global warming is less than a few degrees Celsius, climate change may have a generally positive impact on food production. If warming is greater than a few degrees, the effects will be predominantly negative. In tropical regions and especially in Africa, change is expected to be negative, even at the lower end of the predicted warming.

In Europe, agricultural systems are very advanced and are expected to be able to adapt to climate change. There is, for instance, potential for north and eastward expansion of crops such as maize and sunflowers, which are vulnerable to adverse weather conditions, particularly low soil temperatures. In the UK the growth of maize is currently close to its environmental limits, but higher temperatures will make it easier to grow in this part of the world. Meanwhile, increasing temperatures may have a detrimental effect on other crops such as potatoes, which are sensitive to hot, dry summers. A significant and increasing proportion of the UK's root and vegetable harvest is produced using irrigation, partly in response to supermarket demand for a consistent high quality product, but also in response to the more frequent occurrence of dry growing seasons in recent years.

Agricultural activities currently contribute significant amounts of greenhouse gases to the atmosphere: carbon dioxide *via* deforestation and burning; methane *via* livestock and rice production; and nitrous oxide *via* use of fertilizers and pesticides. Greenhouse gas emissions from agricultural sources to account for some 25% of today's anthropogenic greenhouse gas emissions. These emissions are only likely to increase in the future, given the necessity to expand food production in order to provide for the world's growing population. The challenge is to continue improving yields while at the same time adapting to climate change and reducing, or at least, stabilizing greenhouse gas emissions.

Increases in the carbon dioxide concentrations are expected to increase the biomass in many crop species (particularly C3 plants such as soybean and rice, in comparison to C4 plants such as corn and sorghum). However, as seen in the greenhouse production of horticultural crops such as lettuce, increases in biomass may not always translate to acceptable nutritional quality of the consumed product, due to the increases in the accumulation of starch in the foliage at the expense of soluble carbohydrates and decreases in the protein content, unless nitrogen fertilization is practiced. In as much as the beneficial effects of increasing carbon dioxide concentrations on crop production are frequently cited, increasing levels of surface ozone (generated from precursor pollutant emissions, primarily from fossil fuel combustion) and ultraviolet-B radiation produce deleterious effects. Plant canopies serve as a sink for carbon dioxide during photosynthesis (daytime), they are a source during the night due to respiration. Furthermore, soils are a larger source for carbon dioxide due to increased microbial respiration and litter or biomass decomposition under increasing temperatures. Equally importantly, agricultural emissions of greenhouse gases are governed by the use of nitrogen fertilizers (increased nitrous oxide emissions from the soil as a greenhouse gas) and animal husbandry (increased ammonia emissions from manure leading to atmospheric nitrogen fertilization by deposition). These are major concerns in the context of the contribution of agriculture to the climate change issue. In addition, rice paddies are a major source of methane (a greenhouse gas). As populations grow in Asia (by some 29% in 2030), so will the area under rice production, to sustain the food supply.

Ecosystems and Biodiversity

Climate is an integral part of ecosystems and organisms have adapted to their regional climate over time. Climate change is a factor that has the potential to alter ecosystems and the many resources and services they provide to each other and to society. Human societies depend on ecosystems for the natural, cultural, spiritual, recreational and aesthetic resources they provide.

Ecosystems are becoming more vulnerable and their long-term capacity to adapt is decreasing drastically. Changes in the climate also impact biological diversity and thereby an ecosystem's ability to deliver goods and services for human well-being. Continued increases in greenhouse gas concentrations in the atmosphere from the combustion of fossil fuels could trigger large-scale changes in global biodiversity and require thousands of years of recovery. Climate change is altering the pattern of life on the planet, causing widespread species extinction, migration and behaviour changes. A changing climate forces plants and animals to migrate in order to survive. To make matters worse, human settlements and infrastructure have already subdivided ecosystem habitat into isolated patches. Climate change will make many of these patches uninhabitable for the species that live there, and they will be unable to escape.

Many plant and animal species are unlikely to survive climate change. For some of these species there will no longer be anywhere suitable to live. Others will be unable to reach places where the climate is suitable. A rapid shift to technologies that do not produce greenhouse gases, combined with carbon sequestration, could save 15–20% of species from extinction. The effects of climate change on biodiversity are far-reaching and operate at many different levels - from individuals to ecosystems. At the species level, climate change affects particular species in different ways. It may alter their distribution, abundance, behaviour, phenology (the timing of events such as migration or breeding), morphology (size and shape) and genetic composition. Distributions of plants and animals are heavily influenced by temperature and moisture patterns. Fossil and pollen distribution show that species are very sensitive to climate changes. During several of the Pleistocene interglacial periods, the temperature was 2 to 3 degrees higher and vegetation was drastically different. When the climate changes, species often die out in their present areas and colonize new areas. Therefore, as the climate changes in the future, there will be disruption of natural communities and extinction of populations and species. Some species will be able to disperse to new areas but many will not be able to colonize new areas in sufficient time. A rapid increase of 3 degrees may exceed the ability of many species to adapt.

The Arctic and Antarctic aquatic ecosystems are among the most productive seas on earth. They provide an important source of food for humans as well as endemic and migratory species. As the greatest overall temperature changes are predicted to occur in these important regions, particularly the Arctic, the consequences will be severe. A substantial decrease in coverage and thickness of Arctic sea ice with global warming has been reported. The Antarctic is also experiencing measurable warming and huge sections of ice shelf have collapsed introducing large amounts of freshwater into the marine environment. This depletion of ice coverage could further exacerbate the effects of climate change as there will be less reflection of heat back into the atmosphere. The present trend in ice loss favours the lifestyle of open water species as many seabirds and marine mammals are losing areas of sea ice essential for reproduction and feeding. Furthermore, in these high latitudes the option for many species to move to cooler areas is not available. Climate change will, therefore, have significant implications for the distribution and abundance of polar marine species.

Polar bears have long captured the attention of the general public but probably at no time in the past have they been more in the forefront of the public's imagination than today. Today's heightened interest in polar bears may be due in part to an enhanced understanding of the ecology of polar bears, their environment, and an increased interest in Arctic issues brought on by concerns for climate change. Polar bears do not occur in large numbers; their ability to replace individuals in the population is very limited and population growth is extremely slow; they are long-lived creatures, which help to offset the low reproductive potential; their populations fluctuate in response to natural factors such as climate and prey availability; and, populations can also be impacted by humans through factors such as hunting, oil spills, shipping, and other activities. The level of human presence and activity in the Arctic continues to build.

The Arctic's climate is changing, with a noticeable warming trend that is affecting polar bears. The region is experiencing the warmest air temperatures in four centuries. Polar bears depend on a frozen platform from which to hunt seals, the mainstay of their diet. Without ice, the bears are unable to reach their prey. In a warming Arctic, the bears may not be able to swim the distances required to reach solid ice if reduced ice coverage leads to more open water. Because polar bears are a top predator in the Arctic, changes in their distribution or numbers could affect the entire Arctic ecosystem.

Polar bears and their prey have evolved to living in the extreme conditions of the Arctic. Polar bears and seals are dependent on sea-ice for foraging, resting, and reproduction. The Arctic ecosystem was shaped by climate and continues to be driven today by climate. Polar bears and ice seals, primarily ringed seals, serve as key indicators of the effects of climate change on the Arctic environment. Today, polar bear populations are facing threats previously unprecedented during recorded history in the Arctic. Climate changes on prey species will have a negative effect on polar bears -

- *increased snow can result in reduced success in successfully entering seal birth lairs*
- *decreased snow or increased seasonal rain patterns could effect seal pupping by not providing adequate snow for construction of birth lairs or if rain fall by collapsing birth lairs thus reducing seal productivity*
- *prey reductions could effect polar bear condition and ultimately cub production and survival*

➢ Changes that alter the period of ice coverage could affect distribution and impact on polar bears -

- *they may spend greater amounts of time on land*
- *extended use of terrestrial areas would ultimately effect physical condition of bears when forced to rely on fat stores*

   - *decreased physical condition could effect production and survival*
   - *using deteriorating pack ice may experience increased energetic costs associated with movements and swimming.*
- Denning could be impacted by unusual warm spells
   - *access to high quality denning areas may be limited or restricted*
   - *use of less desirable denning habitat could have impacts on reproduction and survival*
   - *rain or warming could directly cause snow dens to collapse or be opened to ambient conditions*
   - *loss of thermal insulative properties in opened dens could affect litter survival*

The response to climate warming in animal populations has penetrated to the level of the gene. It affects development, reproduction and dormancy and this response is occurring in diverse groups of animals from insects to birds and mammals.

Chromosomal maps are being developed to allow researchers to narrow their focus to identify specific genes that control the seasonal development of animals. The map shows regions of chromosomes that activate and are apparently evolving in animals in response to climate change. Such information will help predict which animals may survive in changing climates and identify which disease carrying vectors may move northward allowing for the production of appropriate vaccines. Plants and animals are not confronting the stress resulting from climate change directly but rather they are flowering, reproducing and going dormant at different times of the year than they used to. Many species will be unable to change quickly enough and will become extinct.

The chromosomal map for the mosquito, *Wyeomyia smithii* which develop within the carnivorous leaves of pitcher plants has been developed. The mosquito has changed genetically in response to recent rapid climate change and now uses shorter, more southern day lengths to initiate dormancy.

Coastal Zones and Sea Level Rise

Sea level rise can be defined as an increase in sea level caused by global warming through two main processes: the thermal expansion of sea water and the rapid melting of land ice. Over the last 100 years, the global sea level has risen by about 10 to 25 cm. Much of the rise in sea level has been related to the concurrent rise in global temperature over the last 100 years. On this time scale, the warming and the consequent thermal expansion of the oceans may account for about 2-7 cm of the observed sea level rise, while the retreat of glaciers and ice caps may account for about 2-5 cm. One of the most significant potential impacts of climate change is sea level rise that may cause inundation of coastal areas and islands, shoreline erosion, and destruction of important ecosystems such as wetlands and mangroves. As global temperatures increase, sea level rise already underway is expected to accelerate due to a thermal expansion of upper layers of the ocean and melting of glaciers. Rising sea levels inundate wetlands and other low-lying lands, erode beaches, intensify flooding, and increase the salinity of rivers, bays, and groundwater tables. Some of these effects may be further compounded by other effects of a changing climate. A third of the world's population living within 50 km of the coasts and a great proportion and a large proportion of them live much closer to the shoreline. Even a modest sea level rise could inundate low-lying regions, accelerate coastal erosion and force the relocation of communities and infrastructures. Sea level rise is also expected to make coastal cities more vulnerable to extreme weather (such as hurricanes) as well as to destroy important ecosystems such as wetlands and mangroves. Rising sea level inundates low-lying lands, erodes shorelines, exacerbates flooding, and increases the salinity of estuaries and aquifers. Wetlands provide habitat for many species, play a key role in nutrient uptake, serve as the basis for many communities' economic livelihoods, provide recreational opportunities, and protect local areas from flooding. Islands are particularly exposed as many are gradually facing the loss of their fresh water supply due to salt-water intrusion. Low-lying coastal areas and deltas (the land where rivers feed into the ocean) house at least 300 million people and are particularly threatened by sea level rise. Some developing countries are especially exposed to sea level rise due to their low-lying nature and limited financial resources to respond. Among the most vulnerable are countries with large populations in deltaic coastal regions such as Bangladesh, Vietnam, China and Egypt.

Energy Production

Energy production and use are sensitive to changes in the climate. For example, increasing temperatures will reduce consumption of energy for heating but increase energy used for cooling buildings. Climate change effects on energy supply and demand will depend not only on climatic factors, but also on patterns of economic growth, land use, population growth and distribution, technological change and social and cultural trends that shape individual and institutional actions.

Hydropower generation is the energy source that is likely to be most directly affected by climate change because it is sensitive to the amount, timing and geographical pattern of precipitation and temperature. Infrastructure for energy production, transmission and distribution could be affected by climate change. For example, if a warmer climate is characterized by more extreme weather events such as windstorms, ice storms, floods, tornadoes and hail, the transmission systems of electric utilities may experience a higher rate of failure, with attendant costs. Power plant operations can be affected by extreme heat waves. For example, intake water that is normally used to cool power plants become warm enough during extreme heat events that it compromises power plant operations. Finally, some renewable sources of energy could be affected by climate change, although these changes are very difficult to predict. If climate change leads to increased cloudiness, solar energy production could be reduced. Wind energy production would be reduced if wind speeds increase above or fall below the acceptable operating range of the technology. Changes in growing conditions could affect biomass production, a transportation and power plant fuel source that is starting to receive more attention. While climate change dominates as the most significant indirect impact on biodiversity from energy production and consumption, direct impacts associated with energy systems are also important - such as habitat loss and fragmentation from energy infrastructure and production and species impacts from pollution and collision.

Public Lands, Recreational Opportunities and Natural Resources

National Parks, National Wildlife Refuges, and other protected areas harbor unique environments and wildlife not found elsewhere. This raises particular concerns about the vulnerability of these ecosystems to a changing climate. Many parks and refuges are designated to protect rare natural features or particular species of plants and animals. Changes in climate could create new stresses on natural communities, and, in the absence of adaptation, lead to the loss of valued resources. National parks and other protected areas are currently susceptible to events influenced by climatic variability, such as drought, wildfires, impaired air quality, and severe storms. Climate change may change the frequency and severity of these kinds of events. In some regions, the risk for drought and wildfire, for example, may increase with climate change. Along coastal regions, sea level rise could erode and inundate the beaches and the wetlands, precipitating loss of beaches, loss of habitat in estuarine ecosystems, and damage to property and natural resources from storm surges.

Outdoor Recreation and Tourism

Outdoor recreation and tourism are sensitive to changes in temperature, rainfall, snowfall, and storm events, and are thus sensitive to climatic variability and change. Tourism businesses, which usually are location-specific, have a lower potential than tourists themselves to adapt to climate change. Shifts in temperature and precipitation patterns could lead to shifts in a variety of outdoor tourism and recreation opportunities, such as skiing, fishing and hunting. The effects of climate change on tourism in any particular area depend in part on whether the tourist activity is summer or winter-oriented and, for the latter, the elevation of the area and the impact of climate on alternative activities and destinations. The extent to which such ecological changes will affect tourism in parks and other protected areas is uncertain.

Tourism is an industry of primary importance for the world economy. For some countries, tourism is the first source of income and foreign currency, and many local economies heavily depend on tourism. Tourists are sensitive to climate and to climate change, which will affect the relative attractiveness of destinations and hence the motive for international tourists to leave their country of origin. But climate change will impose new challenges to the tourism industry. Unlike natural disasters or terrorist attacks, this is not just a short-term effect that can then be quickly forgotten. Rather, climate change will permanently alter the attraction of some holiday regions.

Tourism is a climate-dependent industry, and many destinations owe their popularity to their pleasant climates during traditional holiday seasons. The locations of climatically ideal tourism conditions are likely to shift pole-ward under projected climate change. Tourists from temperate nations that currently dominate international travel would adapt and take advantage of "new climatic opportunities closer to home". Although the overall worldwide growth in tourism is unlikely to suffer much, it could lead to more tourism spending in temperate nations and less in warmer or tropical destinations that are currently popular.

Climate change affects the eco-destinations in several ways. It could modify or affect different ecosystems which attract tourists from different parts of the world. As temperate regions are getting warmer, the flow of tourists from these regions to tropical areas may be reduced and the tourists may prefer to move pole-ward for recreation. Airline industry may suffer from reduced air traffic from the developed world to developing world. The climate change may also be used as climate-tourism in order to attract tourists to visit the places where changes are taking place in different eco-destinations.

Hydrologic Cycle

Our planet is unique because of the abundance of liquid water which provided the environment for the emergence of life and the development and maintenance of human society. If the earth were five per cent closer to the sun, water could only exist in its hydrosphere as water vapour; if the earth were five per cent further from the sun, water could only exist in its hydrosphere in frozen form. The earth in its present position ensures the existence of a vigorous hydrological cycle involving water vapour, liquid water, and frozen water.

The hydrologic cycle can be thought of as a series of reservoirs, or storage areas, and a set of processes that cause water to move between those reservoirs. The largest reservoir by far is the oceans, which hold about 97% of the earth's water. The remaining 3% is the freshwater so important to our survival, but about 78% of that is stored in ice in Antarctica and Greenland. About 21% of freshwater on the earth is groundwater, stored in sediments and rocks below the surface of the earth. The freshwater in rivers, streams, lakes, and rain is less than 1% of the freshwater on the earth and less than 0.1% of all the water on the earth.

Water moves constantly between these reservoirs through the processes of evaporation, condensation and precipitation, surface and underground flow, and others. The driving force for the hydrologic cycle is the sun, which provides the energy needed for evaporation just as the flame of a gas stove provides the energy necessary to boil water and create steam. Water changes from a liquid state to a gaseous state as it evaporates from the oceans, lakes, streams, and soil. Because the oceans are the largest reservoir of liquid water, that is where most evaporation occurs. The amount of water vapor in the air varies widely over time and from place to place; these variations are felt as humidity. Once water vapor is in the air, it circulates within the atmosphere. When an air package rises and cools, the water vapor condenses back to liquid water around particulates like dust, called condensation nuclei. Initially these condensation droplets are much smaller than raindrops and are not heavy enough to fall as precipitation. These tiny water droplets create clouds. As the droplets continue to circulate within the clouds, they collide and form larger droplets, which eventually become heavy enough to fall as rain, snow, or hail. Though the amount of precipitation varies widely over the surface of the earth, evaporation and precipitation are globally balanced. In other words, if evaporation increases, precipitation also increases; rising global temperature is one factor that can cause a worldwide increase in evaporation from the world's oceans, leading to higher overall precipitation.

Since oceans cover around 70% of the earth's surface, most precipitation falls right back into the ocean and the cycle begins again. A portion of precipitation falls on land, however, and it takes one of several paths through the hydrologic cycle. Some water is taken up by soil and plants, some runs off into streams and lakes, some percolates into the groundwater reservoir, some falls on glaciers and accumulates as glacial ice.

The amount of precipitation that soaks into the soil depends on several factors: the amount and intensity of the precipitation, the prior condition of the soil, the slope of the landscape, and the presence of vegetation. These factors can interact in sometimes surprising ways - a very intense rainfall onto very dry soil, typical of the desert south-west, often will not soak into the ground at all, creating flash-flood conditions. Water that does soak in becomes available to plants. Plants take up water through their root systems; the water is then pulled up through all parts of the plant and evaporates from the surface of the leaves, a process called transpiration. Water

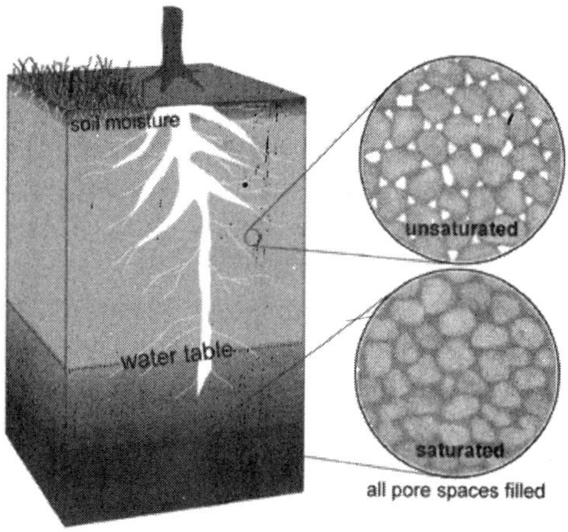

that soaks into the soil can also continue to percolate down through the soil profile into groundwater reservoirs, called aquifers. Aquifers are often mistakenly visualized as great underground lakes; in reality, groundwater fills the pore spaces within sediments or rocks.

Water that does not soak into the soil collects and moves across the surface as run-off, eventually flowing into streams and rivers to get back to the ocean. Precipitation that falls as snow in glacial regions takes a somewhat different journey through the water cycle, accumulating at the head of glaciers and causing them to flow slowly down valleys.

The properties of water and the hydrologic cycle are largely responsible for the circulation patterns in the atmosphere and the oceans on the earth. Atmospheric and oceanic circulations are the major factors that determine the distribution of climatic zones over the earth. Changes in the cycle or circulation can result in major climatic shifts.

Climate change is altering hydrological cycles, with long-term implications for global food availability and the viability of ecosystems. Changes in the frequency and intensity of drought and flooding are already occurring. In addition, mountain snowmelt is coming earlier due to warming temperatures, limiting water supply during peak demand season. Water circulation changes through hydrologic cycle would affect rivers, lakes and all other surface water bodies. Further, they would affect forests, animals, agricultural productivity, cultural practices and eventually humanity.

Atmospheric/Oceanic Circulation

El Nino and La Nina are the most powerful phenomena on the earth and alter the climate across more than half the planet. El Nino-Southern Oscillation commonly referred to as simply El Nino is a global coupled ocean-atmosphere phenomenon. The Pacific Ocean signatures, El Nino and La Nina are important temperature fluctuations in surface waters of the tropical Eastern Pacific Ocean. The name El Niño, from the Spanish for "the little boy", refers to the Christ Child, because the phenomenon is usually noticed around Christ time in the Pacific Ocean off the west coast of South America. El Nino is a shift in ocean temperatures and atmospheric conditions in the tropical Pacific that disrupts weather around the world. It is a poorly understood recurrent climatic phenomenon that primarily affects the Pacific coast of South America, but has dramatic impacts on weather patterns all over the world. Normally, trade winds blow towards the west, across the Pacific, pushing warm surface water away from the South American coast towards Australia and The Philippines. Along the Peruvian coast the water is cold and nutrient-rich, supporting high levels of primary productivity, diverse marine ecosystems, and major fisheries. During El Nino, the trade winds relax in the central and western Pacific. This allows warm water to accumulate in the surface, which causes the nutrients produced by the upwelling of cold water to significantly come down, leading to the killing of plankton and other aquatic life such as fish and the starvation of many seabirds. This is called the El-Nino effect, which is also responsible for destructive disruptions of worldwide weather patterns. In the Indian Ocean, it affects the movement of the monsoon winds which in turn affects the rainfall in the region. The first signs of an El Nino are:

1. Rise in air pressure over the Indian Ocean, Indonesia, and Australia,
2. Fall in air pressure over Tahiti and the rest of the central and eastern Pacific Ocean,
3. Trade winds in the south Pacific weaken or head east,
4. Warm air rises near Peru, causing rain in the northern Peruvian deserts,
5. Warm water spreads from the west Pacific and the Indian Ocean to the east Pacific. It takes the rain with it, causing extensive drought in the western Pacific and rainfall in the normally dry eastern Pacific.

La Nina is the name for the cold phase of ENSO, during which the cold pool in the eastern Pacific intensifies and the trade winds strengthen. The name La Nina originates from Spanish, meaning "the little girl", analogous to El Nino meaning "the little boy". It has also in the past been called anti-El Nino. This is simply 'a cold event' or 'a cold episode'. It is characterized by unusually cold ocean temperatures in the Equatorial Pacific, compared to El Nino, which is characterized by unusually warm ocean temperatures in the Equatorial Pacific. Typically, La Nina occurs roughly half as often as El Nino. The La Nina condition often follows the El Nino, especially when the latter is strong. La Nina causes mostly the opposite effects of El Nino, for example, El Nino would cause a wet period in the Midwestern U.S., while La Nina would typically cause a dry period in this area. This naturally-occurring shift in Pacific weather patterns prompts changes across the globe, causing droughts and flooding in regions as far apart as Africa and India. During La Nina, currents bring nutrients up from the deep water, providing feast, rather than famine, for marine organisms. And accompanying strong winds blow moisture away, making for cloudless skies and dry conditions in equatorial countries from the International Date Line east to South America.

The increased intensity and frequency—now every two to three years—of El Nino and La Nina events in recent decades is due to warmer ocean temperatures resulting from global warming. There are some scientists who believe these phenomena are not related global warming. However, the frequent recurrence of El Nino and La Nina may be related to continued and sustained global warming due to various human activities. If these events happen frequently, the climatic patterns in different regions are most likely to change and accordingly there would be changes in rainfall and temperature regimes throughout the world.

Food Security

Food security is a "situation that exists when all people, at all times, have physical, social, and economic access to sufficient, safe, and nutritious food that meets their dietary needs and food preferences for an active and healthy life". It comprises four key dimensions of food supplies: availability, stability, access, and utilization. The first dimension relates to the availability of sufficient food, i.e., to the overall ability of the agricultural system to meet food demand. Its sub-dimensions include the agro-climatic fundamentals of crop and pasture production and the entire range of socio-economic and cultural factors that determine where and how farmers perform in response to markets. The second dimension, stability, relates to individuals who are at high risk of temporarily or permanently losing their access to the resources needed to consume adequate food, either because these individuals cannot ensure *ex ante* against income shocks or they lack enough "reserves" to smooth consumption *ex post* or both. An important cause of unstable access is climate variability, e.g., landless agricultural laborers, who almost wholly depend on agricultural wages in a region of erratic rainfall and have few savings, would be at high risk of losing their access to food.

Over the past fifty years, human ingenuity has led to technological advances in agriculture that have dramatically increased crop yields. However, despite these improvements, agriculture is still highly dependent on climate since solar radiation, temperature, and precipitation are the main drivers of crop growth. Since the industrial revolution, humans have been changing the global climate by emitting high amounts of greenhouse gases into the atmosphere, potentially resulting in higher global temperatures, changed hydrological regimes, and increased climatic variability.

Climate change over the next century may have significant effects on food supply, i.e., how much food is produced, as well as food security, i.e. how much food is available to people. How much, where, and when food supply and security will be affected by climate change are questions to be solved by many scientists and policy-makers. Overall, climate change, including global warming and increased climate variability, could result in a variety of impacts on agriculture. Some of these effects are biophysical, some are ecological, and some are economic. They include shift in climate and agricultural zones towards the poles, changes in production patterns due to higher temperatures, a boost in agricultural productivity due to increased carbon dioxide in the atmosphere, changing precipitation patterns and increased vulnerability of the landless and the poor.

Climate change affects agriculture and food production in complex ways. It affects food production directly through changes in agro-ecological conditions and indirectly by affecting growth and distribution of incomes, and thus demand for agricultural produce. Changes in temperature and precipitation associated with continued emissions of greenhouse gases will bring changes in land suitability and crop yields. Global and regional weather conditions are also expected to become more variable than at present, with increases in the frequency and severity of extreme events such as cyclones, floods, hailstorms, and droughts. By bringing greater fluctuations in crop yields and local food supplies and higher risks of landslides and erosion damage, they can adversely affect the stability of food supplies and thus food security. Neither climate change nor short-term climate variability and associated adaptation are new phenomena in agriculture. The main concern about climate change and food security is that changing climatic conditions can initiate a vicious circle where infectious disease causes or compounds hunger, which, in turn, makes the affected populations more susceptible to infectious disease. The result can be a substantial decline in labor productivity and an increase in poverty and even mortality. Essentially all manifestations of climate change, be they drought, higher temperatures, or heavy rainfalls have an impact on the disease pressure, and there is growing evidence that these changes affect food safety and food security. Warmer seas may contribute to increased cases of human shellfish and reef-fish poisoning in tropical regions and a pole-ward expansion of the disease.

Economics of Climate Change

The economics of global warming refers to the projected size and distribution of the economic costs and benefits of global warming, and to the economic impacts of actions aimed at the mitigation of global warming. Countries in the polar region are likely to receive large benefits from warming, countries in the mid-latitudes will at first benefit and only begin to be harmed if temperatures rise above 2.5°C. Only countries in the tropical and subtropical regions are likely to be harmed immediately by warming. An industry very directly affected by the risks is the insurance industry; the number of major natural disasters has trebled since the 1960s, and insured losses increased fifteen folds in real terms (adjusted for inflation); 35–40% of the worst catastrophes have been climate change related. Over the past three decades, the proportion of the global population affected by weather-related disasters has doubled.

Rising temperatures and the associate climate change have two opposing direct effects on mortality: higher temperatures in winter reduce deaths from cold; higher temperatures in summer increase heat-related deaths. Global warming is expected to extend the favorable zones for vectors conveying infectious disease such as malaria. In poorer countries, this may simply lead to higher incidence of such diseases. In richer countries, where such diseases have been eliminated or kept in check by vaccination, draining swamps and using pesticides, the consequences may be felt more in economic than health terms, if greater spending on preventative measures is required.

Reducing of greenhouse gas emissions depends on lowering consumption of fossil fuels. The key challenge is that nearly all forms of economic activity rely on fossil fuel energy sources, from transportation fuel, electricity from coal-fired plants, industrial furnaces to home and office heating. Reducing emissions can be achieved through gains in efficiency - producing the same benefits with smaller amounts of fossil energy, or by displacing fossil sources with non- or low-emitting sources. Low emission renewable energy resources such as wind, solar and biomass still represent only a small fraction of total energy consumption. The scale of current fossil energy dependence poses a substantial challenge. Gaining energy efficiency typically requires up-front investment, such as adding insulation, replacing energy-inefficient devices and processes, or buying hybrid vehicles. Some such investments can pay for themselves in the savings on energy bills, and the economic case for choosing them depends on the payback period. Advocates of mitigating climate change hold that greenhouse gas emissions must carry a price, so the market can price in the impact of their emission. This could take the form of a carbon tax or of emission caps, with a market created for trading emission permits, much as was done for sulfate emissions blamed for acid rain. Thus the economic impact of avoiding greenhouse gas emissions depends on how much consumption will have to be avoided, and how quickly the economy can incorporate efficiency gains.

Climate change could cost the global economy almost $7 trillion by 2050 equal to a 20% fall in growth if no action is taken on greenhouse gas emissions. To stabilize levels of carbon dioxide in atmosphere, emissions must start to fall now. Any change in effect will take decades. Expected problems are evacuation of up to 100 million refugees from low-lying areas in flooded areas; water shortages for 1 in 6 persons displaced; fall in crop yields; spread of tropical diseases; insecurity of supplies; and extinction of up to 40% of species.

Vulnerability to climate change will mainly depend on economic position and infrastructure capacity of nations. Climate change effects will impose significant additional stress on ecological and socio-economic systems. Technologically advanced countries are prepared well for responding to climate change, particularly by developing and establishing suitable policy, institutional and social capable for dealing with the consequences. But the poor and developing countries are mostly affected by climate change, because they are not having enough and sound technologies or scientific development to deal with this impact.

In developing countries like India, climate change is an additional burden because ecological and socio-economic systems are already facing pressures from rapid population, industrialization and economic development. India's climate could become warmer under conditions of increased atmospheric carbon dioxide. The average temperature change is predicted to be in the range of 2.33°C to 4.78°C with the doubling in CO_2 concentrations. In India, climate change can still seem an esoteric issue, confined to the bastions of elite research institutions and a handful of bureaucrats. The slow, insidious changes it will bring over the next century are not yet on the radar of everyday people, or even policymakers. The implications have simply not yet captured media attention in the same way as major natural disasters.

4

Ozone Layer

The Earth's atmosphere is a layer of gases that surround the planet Earth and retained by the Earth's gravity. It contains roughly by molar content/volume 78% nitrogen, 21% oxygen, 1% argon, 0.38% carbon dioxide, trace amounts of other gases and a variable amount of water vapour. These mixture of gases are commonly known as air. The atmosphere protects life on Earth by absorbing ultraviolet solar radiation and reducing temperature extremes between day and night. There is no definite boundary between the atmosphere and outer space. It slowly becomes thinner and fades into space. These quarters of the atmosphere's mass is within 11 km of the planetary surface. The boundary between atmosphere and outer space is at an altitude of 100 km. The temperature of the earth's atmosphere varies with altitude; the mathematical relationship between temperature and altitude varies among five different atmosphere layers.

Exosphere

The exosphere is the highest layer of the atmosphere. Together with the ionosphere, it makes up the thermosphere. The exosphere extends to 10,000 km above the Earth's surface. This is the upper limit of our atmosphere. The atmosphere here merges into space in the extremely thin air. Air atoms and molecules are constantly escaping to space from the exosphere. In this region of the atmosphere, hydrogen and helium are the prime components and are only present at extremely low densities. This is the area where many satellites orbit the Earth.

Ionosphere

The ionosphere is a name for the layer of the earth's atmosphere that is ionized by solar radiation. The more common name for this radiation is solar wind. Even though many believe the space around earth is a vacuum, it is not completely empty. The sun's upper atmosphere (the corona) is very hot and some of its hydrogen and helium are able to escape from the sun's gravity. Because the gas is hot and is in a constant stream of solar energy and it becomes fully ionized plasma. This streaming plasma is the solar wind, and it flows out past the earth affecting the earth's magnetic field, the magnetosphere and ionosphere. The Earth receives a lot of energy from the sun in the form of radiation- about 1370 Watts per square meter. Longer wavelength radio signals can be "bounced" off the ionosphere allowing radio communication "over the horizon". This is how the long, medium and short wave radio broadcasts reach receivers over long distances. Because the ionosphere is not a nice smooth "mirror" the signal can be scattered in many directions causing loss of signal strength and interference from other transmitters. The ionosphere is particularly disturbed in the auroral regions, and during magnetic sub-storms. Shorter wavelength radio signals pass through the ionosphere but are affected by it. These shorter wavelengths are used by satellites for imaging the earth, and the ionosphere affects the images rather like the way the atmosphere causes "twinkling" of the stars. The ionosphere is important for radio, television and satellite communications.

Thermosphere

The thermosphere or heat sphere is the outer layer of the atmosphere, separated from the mesosphere by the mesopause. It is the fourth layer of the Earth's atmosphere and is located above the mesosphere. Within the thermosphere temperatures rise continually to well beyond 1000°C. The few molecules that are present in the thermosphere receive extraordinary amounts of energy from the Sun, causing the layer to warm to such high temperatures. Air temperature, however, is a measure of the kinetic energy of air molecules, not of the total energy stored by the air. Therefore, since the air is so thin within the thermosphere, such temperature values are not comparable to those of the troposphere or stratosphere. Although the measured temperature is very hot, the thermosphere would actually feel very cold to us because the total energy of only a few air molecules residing there would not be enough to transfer any appreciable heat to our skin.

Mesosphere

The mesosphere is the third highest layer in our atmosphere, occupying the region 50 km to 80 km above the surface of the Earth, above the troposphere and stratosphere, and below the thermosphere. It is separated from the stratosphere by the stratopause and from the thermosphere by the mesopause. Temperatures in the mesosphere drop with increasing altitude to about -100°C. The mesosphere is the coldest of the atmospheric layers. In fact it is colder than Antarctica's lowest recorded temperature. It is cold enough to freeze water vapour into ice clouds. One can see these clouds if sunlight hits them after sunset. They are called Noctilucent Clouds. The mesosphere is also the layer in which a lot of meteors burn up while entering the Earth's atmosphere. From the Earth they are seen as shooting stars.

Stratosphere

The stratosphere is the second major layer of the atmosphere. It lies above the troposphere and is separated from it by the tropopause. It occupies the region of atmosphere from about 12 to 50 km, although its lower boundary tends to be higher nearer the equator and lower nearer the poles. The stratosphere defines a layer in which temperature rises with increasing altitude. At the top of the stratosphere the thin air may attain temperatures close to 0°C. This rise in temperature is caused by the absorption of ultraviolet radiation from the Sun by the ozone layer. Such a temperature profile creates very stable atmospheric conditions, and the stratosphere lacks the air turbulence that is so prevalent in the troposphere. Consequently, the stratosphere is almost completely free of clouds or other forms of weather. The stratosphere provides some advantages for long-distant flight because it is above stormy weather and has strong, steady, horizontal winds.

The thin ozone layer in the upper stratosphere has a high concentration of ozone, a particularly reactive form of oxygen. The formation of this layer is a delicate matter, since only when oxygen is produced in the atmosphere can an ozone layer form and prevent an intense flux of ultraviolet radiation from reaching the surface, where it is quite hazardous to the evolution of life. For nearly a billion years, ozone molecules in the atmosphere have protected life on the Earth from the effects of ultraviolet rays. Only 10 or less of every million molecules of air are ozone. The majority of these ozone molecules reside in a layer between 10 and 40 kilometers above the Earth's surface in the stratosphere.

Troposphere

The lowest layer of the atmosphere is called the troposphere. It ranges in thickness from 8 km at the poles to 16 km over the equator. The troposphere is bounded above by the tropopause, a boundary marked by stable temperatures. Above the troposphere is the stratosphere. Although variations do occur, temperature usually declines with increasing altitude in the troposphere. The troposphere is denser than the layers of the atmosphere above it because of the weight compressing it, and it contains roughly 80% of the total mass of the atmosphere; 50% of the total mass of the atmosphere is located in the lower 5 km of the troposphere.

It is primarily composed of nitrogen (78%) and oxygen (21%) with only small concentrations of other trace gases. Nearly all atmospheric water vapour or moisture is found in the troposphere. The troposphere has a great deal of vertical mixing due to solar heating at the surface. This heating warms air masses, which makes them less dense so they rise. When an air mass rises, the pressure upon it decreases so it expands, doing work against the opposing pressure of the surrounding air. To do work is to expend energy, so the temperature of the air mass decreases. As the temperature decreases, water vapor in the air mass may condense or solidify, releasing latent heat that further uplifts the air mass. This process determines the maximum rate of decline of temperature with height, called the adiabatic lapse rate.

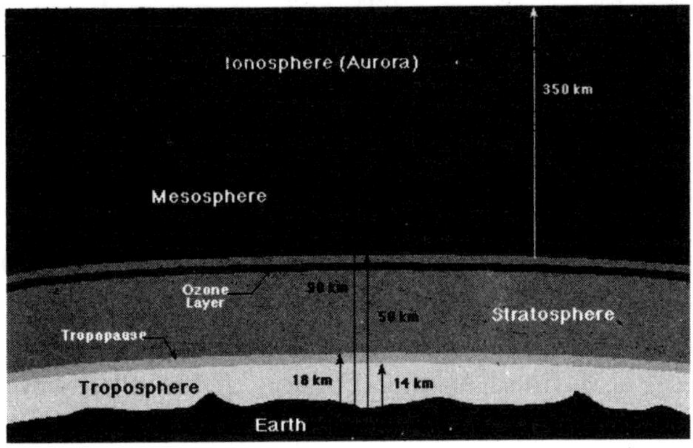

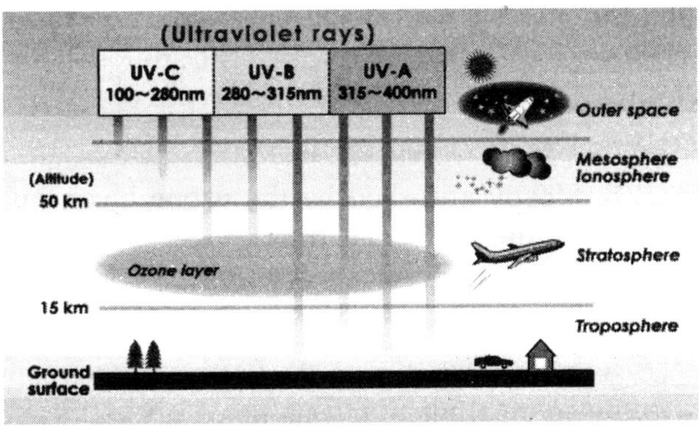

Ultraviolet Radiation

All energies that move at the speed of light are collectively referred to as electromagnetic radiation or 'light'. Various types of light differ in their wavelength, frequency and energy; higher energy waves have higher frequencies and shorter wavelengths. Pigments inside the retina of our eyes absorb wavelengths of light between 400 nm - 700 nm, collectively referred to as 'visible light'. A "nm" is a nanometer which is one billionth meters.

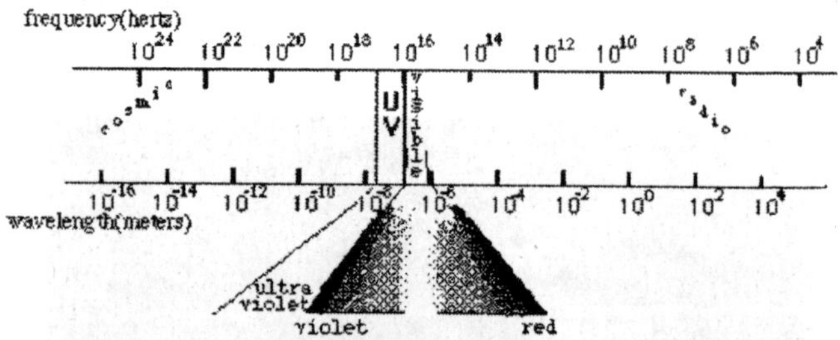

Stratospheric oxygen and ozone molecules absorb 97-99% of the sun's high frequency Ultraviolet light, light with wavelengths between 150 nm and 300 nm. Ultraviolet (UV) radiation is defined as that portion of the electromagnetic spectrum between X-rays and visible light, i.e., between 40 and 400 nm. The UV spectrum is divided into Vacuum UV (40-190 nm), Far UV (190-220 nm), UVC (220-290 nm), UVB (290-320 nm), and UVA (320-400 nm). Absorption levels of UVA, UVB, and UVC by ozone vary. The stratospheric ozone layer absorbs some, but not all, of these types of UV radiation:

- UVA: Not absorbed by the ozone layer.
- UVB: Mostly absorbed by the ozone layer, but some do reach the Earth's surface.
- UVC: Completely absorbed by the ozone layer and oxygen.
- UVA and UVB that reach the Earth's surface contribute to the serious health effects listed above.

The sun is our primary natural source of UV radiation. Artificial sources include tanning booths, black lights, curing lamps, germicidal lamps, mercury vapor lamps, halogen lights, high-intensity discharge lamps, fluorescent and incandescent sources, and some types of lasers. Unique hazards apply to the different sources depending on the wavelength range of the emitted UV radiation.

UVC is almost never observed in nature because it is absorbed completely in the atmosphere, as are Far UV and Vacuum UV. Germicidal lamps are designed to emit UVC radiation because of its ability to kill bacteria. In humans, UVC is absorbed in the outer dead layers of the epidermis. Accidental overexposure to UVC can cause corneal burns, commonly termed welders' flash, and snow blindness, a severe sunburn to the face.

The amount of UVB light received by a location is strongly dependent on latitude and elevation of the location, cloud cover and proximity to an industrial area. In industrial areas, the photochemical smog offers protection from UVB radiation. Industrial processes produce ozone which absorbs UVB. UVB is typically the most destructive form of UV radiation because it has enough energy to cause photochemical damage to cellular DNA. UVB is needed by humans for synthesis of vitamin D; however, harmful effects can include erythema (sunburn), cataracts, and development of skin cancer. Individuals working outdoors are at the greatest risk of UVB effects. Most solar UVB is blocked by ozone in the atmosphere, and there is concern that reductions in atmospheric ozone could increase the prevalence of skin cancer.

UVA is the most commonly encountered type of UV light. UVA exposure has an initial pigment-darkening effect (tanning) followed by erythema if the exposure is excessive. Atmospheric ozone absorbs very little of this part of the UV spectrum. UVA is needed by humans for synthesis of vitamin D; however, overexposure to UVA has been associated with toughening of the skin, suppression of the immune system, and cataract formation. UVA light is often called black light.

Human Health Effects of UVB Light

DNA absorbs UVB light and the absorbed energy can break bonds in the DNA. Most of the DNA breakages are repaired by proteins present in the cells nucleus but un-repaired genetic damage of the DNA can lead to skin cancers. The principle danger of skin cancer is to light-skinned peoples. A 1% decrease in the ozone layer will cause an estimated 2% increase in UVB irradiation; it is estimated that this will lead to a 4% increase in basal carcinomas and 6% increase in squamous-cell carcinomas. 90% of the skin carcinomas are attributed to UVB exposure. The above named carcinomas are relatively easy to treat, if detected in time, and are rarely fatal.

Possible eye damage can result from high doses of UV light, particularly to the cornea which is a good absorber of UV light. High doses of UV light can causes a temporary clouding of the cornea, called 'snow-blindness', and chronic doses have been tentatively linked to the formation of cataracts. Higher incidences of cataracts are found at high elevations, Tibet and Bolivia; and higher incidences are seen at lower latitudes (approaching the equator).

The penetration of increased amounts of UVB light has caused great concern over the health of marine plankton that densely populates the top 2 meters of ocean water. The natural protective-response of most chlorophyll containing cells to increased light-radiation is to produce more light-absorbing pigments but this protective response is not triggered by UVB light. Another possible response of plankton is to sink deeper into the water but this reduces the amount of visible light they need for photosynthesis, and thereby reduces their growth and reproduction rate. In other words, the amount of food and oxygen produced by plankton could be reduced by UV exposure without killing individual organisms.

Ozone

It is named from the Greek "ozein" for smell. It was first discovered in 1839 by German scientist Christian Friedrich

Schonbein. It is a pale blue, relatively unstable molecule made up of three oxygen atoms and thus has the formula O_3. It is angular, polar, and diamagnetic. Both oxygen bond lengths (1.28 angstroms) are identical. Only 10 or less of every million molecules of air are ozone. The majority of these ozone molecules reside in a layer between 10 and 40 kilometers above the Earth's surface in the stratosphere.

One billion years ago, early aquatic organisms called blue-green algae began using energy from the Sun to split molecules of H_2O and CO_2 and recombine them into organic compounds and molecular oxygen (O_2). This solar energy conversion process is known as photosynthesis. Some of the photosynthetically created oxygen combined with organic carbon to recreate CO_2 molecules. The remaining oxygen accumulated in the atmosphere, touching off a massive ecological disaster with respect to early existing anaerobic organisms. As oxygen in the atmosphere increased, CO_2 decreased. High in the atmosphere, some oxygen (O_2) molecules absorbed energy from the Sun's ultraviolet (UV) rays and split to form single oxygen atoms. These atoms combined with remaining oxygen (O_2) to form ozone (O_3) molecules, which are very effective at absorbing UV rays. The thin layer of ozone that surrounds Earth acts as a shield, protecting the planet from irradiation by UV light. The amount of ozone required to shield Earth from biologically lethal UV radiation, wavelengths from 200 to 300 nanometers (nm), is believed to have been in existence 600 million years ago. At this time, the oxygen level was approximately 10% of its present atmospheric concentration. Prior to this period, life was restricted to the ocean. The presence of ozone enabled organisms to develop and live on the land. Ozone played a significant role in the evolution of life on Earth, and allows life as we presently know it to exist. Ozone is produced naturally in the stratosphere when highly energetic solar radiation strikes molecules of oxygen, O_2, and cause the two oxygen atoms to split apart in a process called photolysis. If a freed atom collides with another O_2, it joins up, forming ozone

O_3. Most of the ozone in the stratosphere is formed over the equatorial belt, where the level of solar radiation is greatest. The circulation in the atmosphere then transports it towards the pole. So, the amount of stratospheric ozone above a location on the Earth varies naturally with latitude, season, and from day-to-day.

Ozone is formed from molecular oxygen (O_2) by ultraviolet and extreme ultraviolet photolysis followed by recombination of atomic oxygen (O) with O_2. It may also be formed by passing an electrical discharge through gaseous oxygen. It is characterized by a unique odor that is often noticed during electrical storms and in the vicinity of electrical equipment. The density of ozone is about 2.5 times that of O_2. At -112°C, it condenses to a deep blue liquid. It is a powerful oxidizing agent and, as a concentrated gas or a liquid, is highly explosive. Excess oxygen atoms, also known as free radicals, oxidize materials that they contact and are associated with the aging process.

Ozone reacts strongly with other molecules; large concentrations of ozone near the ground prove toxic to living things. Motor vehicle exhaust and industrial emissions, gasoline vapors, and chemical solvents are some of the major sources of NOx and VOC, also known as ozone precursors. Strong sunlight and hot weather cause ground-level ozone to form harmful concentrations in the air. Many urban areas tend to have high levels of "bad" ozone, but other areas are also subject to high ozone levels as winds carry NOx emissions hundreds of miles away from their original sources. At higher altitudes, where 90 per cent of our planet's ozone resides, it does a remarkable job of absorbing ultraviolet radiation. In the absence of this gaseous shield in the stratosphere, the harmful radiation has a perfect portal through which to strike Earth. The amounts of "good" and "bad" ozone in the atmosphere depend on a balance between processes that create ozone and those that destroy it. An upset in the ozone balance can have serious consequences for life on Earth. Ozone is also naturally broken down in the stratosphere. In an unpolluted atmosphere there is a balance between the amount of ozone being produced and destroyed and so the total concentration remains relatively constant. At different temperatures and pressures (i.e. varying altitudes), there are different production and destruction reaction rates leading to a variation in concentration. The highest ozone concentrations are in the lower stratosphere, between about 18 and 26 km.

Ozone also occurs in very small amounts in the troposphere. It is produced at ground level through a reaction between sunlight and, e.g., gases emitted from cars. As a pollutant it should not be confused with the separate problem of stratospheric ozone depletion. The stratospheric ozone layer shields life on Earth from the Sun's harmful ultraviolet radiation. Chemicals that destroy ozone are formed by industrial and natural processes. With the exception of volcanic injection and aircraft exhaust, these chemicals are carried up into the stratosphere by strong upward-moving air currents in the tropics. Methane (CH_4), chlorofluorocarbons (CFCs), nitrous oxide (N_2O) and water are injected into the stratosphere through towering tropical cumulus clouds. These compounds are broken down by the ultraviolet radiation in the stratosphere. Byproducts of the breakdown of these

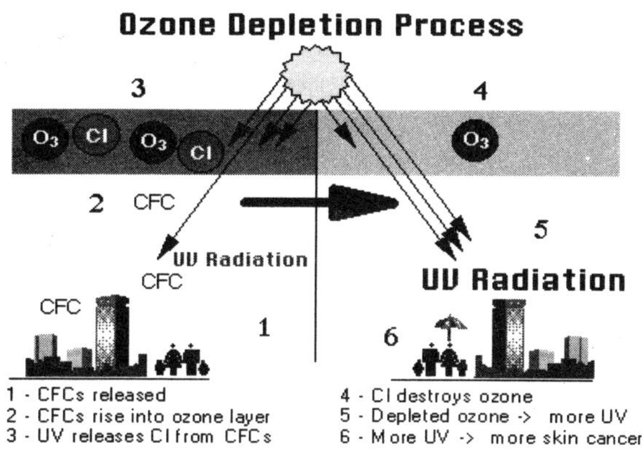

chemicals form "radicals" such as nitrogen dioxide (NO_2) and chlorine monoxide (ClO) that play an active role in ozone destruction. Aerosols and clouds can accelerate ozone loss through reactions on cloud surfaces. Thus, volcanic clouds and polar stratospheric clouds can indirectly contribute to ozone loss. Ozone is destroyed by reactions with chlorine, bromine, nitrogen, hydrogen, and oxygen gases. Reactions with these gases typically occur through catalytic processes. A catalytic reaction cycle is a set of chemical reactions which result in the destruction of many ozone molecules while the molecule that started the reaction is reformed to continue the process. Because of catalytic reactions, an individual chlorine atom can on average destroy nearly a thousand ozone molecules before it is converted into a form harmless to ozone.

Stratospheric ozone shield that screens the earth from some of the sun's harmful ultraviolet rays is depleted by some chemicals that result from various human activities. The most commonly known ozone depletion chemicals are the chlorofluorocarbons, the methyl halides, carbon tetrachloride (CCl_4), carbon tetrafluoride (CF_4), and the halons which contain bromine instead of chlorine. Such compounds are called halocarbons. CFCs have been widely used as refrigerants, solvents, foam blowing agents, as aerosol propellants; halons are used primarily as firefighting agents. Halons are used primarily in fire extinguishers. Halon-1301 has an ozone depleting potential 10 times that of CFC-11. Although the use of halons in developed countries has been phased out since 1996, the atmospheric concentration of these potent, ozone destroyers is still rising because of their long atmospheric lifetimes. To date halons have accounted for about 5% of global ozone depletion. Methyl bromide, another bromine-containing halocarbon, has been used as a pesticide since the 1960s. Human sources of methyl bromide have been responsible for approximately 5 to 10% of global ozone depletion. Increased radiation from ozone depletion in stratosphere could result in an increase in skin cancers, suppression of the human immune system, and decreased productivity of terrestrial and aquatic organisms, including some commercially important crops.

Ozone Cycle

Three forms or allotropes of oxygen are involved in the ozone-oxygen cycle: oxygen atoms (O or atomic oxygen), oxygen gas (O_2 or diatomic oxygen), and ozone gas (O_3 or triatomic oxygen). Ozone is formed in the stratosphere when oxygen molecules photo dissociate after absorbing an ultraviolet photon whose wavelength is shorter than 240 nm. This produces two oxygen atoms. The atomic oxygen then combines with O_2 to create O_3. Ozone molecules absorb UV light between 310 and 200 nm, following which ozone splits into a molecule of O_2 and an oxygen atom. The oxygen atom then joins up with an oxygen molecule to regenerate ozone. This is a continuing process which terminates when an oxygen atom "recombines" with an ozone molecule to make two O_2 molecules: $O + O_3 = 2\ O_2$. The overall amount of ozone in the stratosphere is determined by a balance between photochemical production and recombination.

Ozone can be destroyed by a number of free radical catalysts, the most important of which are the hydroxyl radical (OH), the nitric oxide radical (NO) and atomic chlorine (Cl) and bromine (Br·). All of these have both natural and anthropogenic sources; at the present time, most of the OH· and NO· in the stratosphere is of natural origin, but human activity has dramatically increased oxygen, chlorine and bromine. These elements are found in certain stable organic compounds, especially chlorofluorocarbons (CFCs), which may find their way to the stratosphere without being destroyed in the troposphere due to their low reactivity. Once in the stratosphere, the Cl and Br atoms are liberated from the parent compounds by the action of ultraviolet light, e.g. ('h' is Planck's constant, 'v' is frequency of electromagnetic radiation).

$$CFCl_3 + hv = CFCl_2 + Cl$$

The Cl and Br atoms can then destroy ozone molecules through a variety of catalytic cycles. In the simplest example of such a cycle, a chlorine atom reacts with an ozone molecule, taking an oxygen atom with it (forming ClO) and leaving a normal oxygen molecule. A free oxygen atom then takes away the oxygen from the ClO, and the final result is an oxygen molecule and a chlorine atom, which then reinitiates the cycle. The chemical shorthand for these gas-phase reactions is:

$$Cl + O_3 \rightarrow ClO + O_2$$

$$ClO + O \rightarrow Cl + O_2$$

The net reaction is: $O_3 + O \rightarrow 2\ O_2$, the "recombination" reaction given above. The overall effect is to increase the rate of recombination, leading to an overall decrease in the amount of ozone. For this particular mechanism to operate there must be a source of O atoms, which is primarily the photo dissociation of O_3; thus this mechanism is only important in the upper stratosphere where such atoms are abundant. More complicated mechanisms have been discovered that lead to ozone destruction in the lower stratosphere as well. A single chlorine atom would keep on destroying ozone for up to two years, the time scale for transport back down to the troposphere. On a per atom basis, bromine is even more efficient than chlorine at destroying ozone, but there is much less bromine in the atmosphere at present. As a result, both chlorine and bromine contribute significantly to the overall ozone depletion. Fluorine and iodine atoms participate in analogous catalytic cycles. However, in the Earth's stratosphere, fluorine atoms react rapidly with water and methane to form strongly-bound HF, while organic molecules which contain iodine react so rapidly in the lower atmosphere that they do not reach the stratosphere in significant quantities. Furthermore, a single chlorine atom is able to react with 1,00,000 ozone molecules. This fact plus the amount of chlorine released into the atmosphere by chlorofluorocarbons (CFCs) yearly demonstrates how dangerous CFCs are to the environment.

Ozone Hole

The ozone hole is not technically a "hole" where no ozone is present, but is actually a region of exceptionally depleted ozone in the stratosphere over the Antarctic that happens at the beginning of Southern Hemisphere spring (August-October). Satellite instruments provide us with daily images of ozone over the Antarctic region. The decrease in the ozone layer was predicted in the early 1980's to be roughly 7% over a sixty-year period. The sudden and, at that time unexplained, disappearance of over 50% of the ozone layer in a localized area of Antarctica created quite a stir. Many were worried that ozone holes might start to appear over other areas of the globe but to date the only other significant, localized depletion is a much smaller ozone "dimple" observed during the arctic spring over the North Pole. The Antarctic hole is a warning that if conditions become more Antarctic: cooler stratospheric temperatures, more stratospheric clouds, more active chlorine; then global ozone will decrease at a much greater pace. Some of the more popular senarios of global warming predict that these changes could occur in larger portions of the stratosphere. When the antarctic ozone hole breaks-up, the ozone-depleted air drifts out into nearby areas.

Polar Regions get a much larger variation in sunlight than anywhere else, and during the 3 months of winter spend most of time in the dark without solar radiation. Temperatures hover around or below -80°C for much of the winter and the extremely low Antarctic temperatures cause cloud formation in the relatively "dry" stratosphere. These Polar Stratospheric Clouds are composed of ice crystals that provide the surface for a multitude of reactions, many of which speed up the degradation of ozone molecules.

The ozone hole is a completely different phenomenon to global warming; however, there are links between them. The ozone hole is caused by ozone depleting chemicals in the atmosphere, which have been produced by industry, for example CFCs. One link is that CFCs are also 'greenhouse gasses'. Enhanced global warming is a probable consequence of increasing amounts of 'greenhouse gasses', such as carbon dioxide and methane, in the atmosphere. Although the surface of the earth warms, higher up the atmosphere cools, thus increasing the area where stratospheric clouds can form. This makes a larger area susceptible to ozone depletion and provides another link between the two issues.

The northern and southern hemispheres have different 'weather' in the ozone layer, and the net result is that the temperature of the Arctic ozone layer during winter is normally some ten degrees warmer than that of the Antarctic. This means that such clouds are rare, but sometimes the 'weather' is colder than normal and they do form. Under these circumstances significant ozone depletion can take place over the Arctic, but it is usually for a much shorter period of time and covers a smaller area than in the Antarctic. Antarctica was an early victim of ozone destruction. A massive hole in the ozone layer right above Antarctica now threatens not only that continent, but many others that could be the victims of Antarctica's melting icecaps.

Each spring in the stratosphere over Antarctica (Spring in the southern hemisphere is from September through November), atmospheric ozone is rapidly destroyed by chemical processes. As winter arrives, a vortex of winds develops around the pole and isolates the polar stratosphere. When temperatures drop below -78°C, thin clouds form of ice, nitric acid, and sulphuric acid mixtures. Chemical reactions on the surfaces of ice crystals in the clouds release active forms of CFCs. Ozone depletion begins, and the ozone "hole" appears

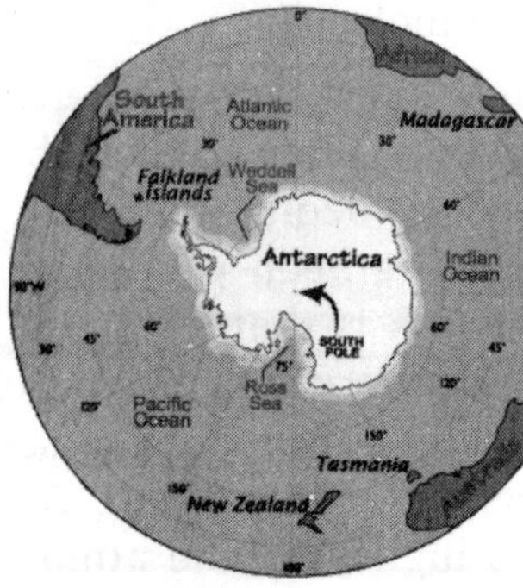

over the course of two to three months, approximately 50% of the total column amount of ozone in the atmosphere disappears. At some levels, the losses approach to 90%. This has come to be called the Antarctic ozone hole. In spring, temperatures begin to rise, the ice evaporates, and the ozone layer starts to recover. The Antarctic ozone hole was discovered in 1985 by British scientists Joesph Farman, Brian Gardiner, and Jonathan Shanklin of the British Antarctic Survey. The ozone "hole" is really a reduction in concentrations of ozone high above the earth in the stratosphere. The ozone hole is defined geographically as the area wherein the total ozone amount is less than 220 Dobson Units. The ozone hole has steadily grown in size (up to 27 million sq. km.) and the length of existence (from August through early December) over the past two decades.

Chlorofluorocarbons (CFCs) are a family of chemical compounds developed back in the 1930's as safe, non-toxic, non-flammable alternative to dangerous substances like ammonia for purposes of refrigeration and spray can propellants. Their usage grew enormously over the years. One of the elements that make up CFCs is chlorine. Very little chlorine exists naturally in the atmosphere. But it turns out that CFCs are an excellent way of introducing chlorine into the ozone layer. The ultraviolet radiation at this altitude breaks down CFCs, freeing the chlorine. Under the proper conditions, this chlorine has the potential to destroy large amounts of ozone. This has indeed been observed, especially over Antarctica. As a consequence, levels of genetically harmful ultraviolet radiation have increased. Chlorofluorocarbons (CFCs) are highly stable compounds that are used as propellants in spray cans and in refrigeration units. There are several organic compounds composed of carbon, fluorine, chlorine, and hydrogen. CFCs are manufactured under the trade name Freon.

Developed during the 1930s, CFCs found wide application after World War II. These halogenated hydrocarbons, notably trichlorofluoromethane (CFC-11, or F-11) and dichlorodifluoromethane (CFC-12, or F-12), have been used extensively as aerosol-spray propellants, refrigerants, solvents, and foam-blowing agents. They are well-suited for these and other applications because they are non-toxic and non-flammable and can be readily converted from a liquid to a gas and vice versa.

Chlorofluorocarbons are non-flammable and non-carcinogenic. They contain fluorine atoms, carbon atoms and chlorine atoms. The 5 main CFCs include CFC-11 (trichlorofluoromethane - $CFCl_3$), CFC-12 (dichloro-difluoromethan-CF_2C-l2), CFC-113 (trichloro-trifluoroethane - $C_2F_3Cl_3$), CFC-114 (dichloro-tetrfluoroethane -$C_2F_4Cl_2$), and CFC-115 (chloropentafluoroethane - C_2F_5Cl). CFCs have a lifetime in the atmosphere of about 20 to 100 years, and consequently one free chlorine atom from a CFC molecule can do a lot of damage, destroying ozone molecules for a long time. Although emissions of CFCs around the developed world have largely ceased due to international control agreements, the damage to the stratospheric ozone layer will continue well into the 21st century.

It is an open secret that ice caps of the Earth are quickly diminishing in size. Two decades ago the ozone hole was regarded as the major threat to the biosphere of the planet. The ozone hole prevents the quick melting of the Antarctic. Antarctica resists the warming better than the Arctic does. The North Pole is getting warmer much quicker than the South Pole. Thus, in the regions of eternal ice in the Northern Hemisphere seasonal temperatures may be a bit more than two degrees Celsius lower than the average, but in the most part of Antarctica there are no changes in temperature. The only exception is the Antarctic Peninsula, a strip of land along South America. There warming is even more measurable than in the Southern Hemisphere - about 1.7 degrees Celsius. A thinner layer of stratospheric ozone at a height of several dozen kilometers above the Antarctic ice causes stronger winds on this continent. It keeps cold air above the mainland and prevents the flow of warmth from the north. While the situation persists, ice will not melt. However, the hole discovered in the mid-1980s is not eternal: the ozone layer is gradually restoring above the South Pole. On the one hand, it helps to protect the Earth from solar radiation, but on the other hand, it means that the last warming-preventing factor will not work. Recovery of the ozone hole above Antarctica could warm the Antarctic and cause more ice to melt in coming decades. As the ozone hole heals, wind patterns that shield the interior of the polar region from warm air may break down, causing warming in the Antarctica as well as warmer and drier conditions in Australia.

International Controls on Ozone Depletion Substances

A strong international consensus that the ozone layer needs to be protected has developed over the past decade. The first step towards turning consensus into global action was taken in March 1985, ahead of firm scientific proof that man-made chemicals were damaging the ozone layer. This was the adoption of the Vienna Convention for the Protection of the Ozone Layer. Parties to the Convention agreed to take 'appropriate measures' to safeguard the ozone layer, and anticipated the negotiation of protocols for specific measures. The need for a protocol arose almost immediately, when the first evidence of the Antarctic ozone hole was published in June 1985. Global negotiations for a protocol were put into top gear, and resulted in adoption in September 1987 of the Montreal Protocol on Substances that Deplete the Ozone Layer. The Montreal Protocol came into force in January 1989 and is the legal basis for the worldwide effort to safeguard the ozone layer through controls on production, consumption and use of ozone-depleting substances. By December 1995, 150 countries had ratified the Montreal Protocol, so becoming Parties to it and legally bound by its requirements. About a third is developed and two-thirds are developing countries. The original Montreal Protocol defined measures that parties had to take to limit production and consumption of eight ozone-depleting substances (ODS), known in the language of the Protocol as 'controlled substances'. At meetings held in

London and Copenhagen in 1990 and 1992, the controls were strengthened and broadened to cover other chemicals. Instead of merely a reduction in production and consumption of five CFCs and three halons, the Protocol now requires developed countries to phase out 15 CFCs, three halons, 34 HBFCs, carbon tetrachloride and methyl chloroform. A longer-term reduction schedule, also leading to complete phase out, has been agreed for 40 HCFCs. The list of controlled substances is now extended to include methyl bromide as agreed at the 7th Meeting of the Parties.

Recognizing developing countries' need for economic development and their relatively low historical use of CFCs, the Montreal Protocol grants developing countries a 'grace period' of ten years more than developed countries to implement the reduction and phase-out measures required by the Protocol. In addition, at their 1990 meeting in London, the Parties created a financial mechanism to provide technical and financial assistance to developing countries' ozone protection programs. To be eligible to receive support under the financial mechanism, Parties must be developing countries and must consume less than 0.3 kg per capita per annum of controlled substances.

The Montreal Protocol is one of the most successful environment protection agreements in the world. It is the first worldwide agreement designed to protect human health and the environment against the adverse effects of the depletion of the stratospheric ozone layer.

Recognizing the strong commitment shown by signatory States and States Parties to the Montreal Protocol, as well as by various concerned non-governmental organizations, the United Nations proclaimed 16 September the International Day for the Preservation of the Ozone Layer, commemorating the date in 1987 on which the Montreal Protocol on Substances that Deplete the Ozone Layer was signed, to be observed beginning in 1995. Since 1995, on 16 September each year, the International Day for the Preservation of the Ozone Layer is celebrated. This commemoration around the world offers an opportunity to focus attention and action at the global, regional and national levels on the protection of the ozone layer. All Member States are invited to devote this special day to promotion, at the national level, of concrete activities in accordance with the objectives and goals of the Montreal Protocol and its Amendment.

The Themes for each year to observe World Ozone Day were as follows: Preservation of Ozone Layer for 1999-2001; Save Our Sky: Protect yourself: Protect the Ozone Layer for 2002; Save Our sky: There is a hole lot more to do for our children for 2003, Save Our Sky: Ozone Friendly Planet, Our Target for 2004, Act Ozone Friendly - Stay Sun Safe for 2005, Protect the Ozone Layer: Save Life on Earth for 2006, Celebrating 20 years of Progress in Ozone Recovery for 2007. In 2007, International Ozone Day was celebrated by the United Nations in the city of Montreal, Canada, where the Protocol was first signed. The summary of UN Secretary-General's message is as follows:

"Twenty years later, the developed world has nearly phased out these substances. And their use in the developing world has plummeted by over 80 per cent. I congratulate all Parties to the Montreal Protocol for this remarkable achievement. Their vision and commitment have translated into real gains."

Therefore, the Montreal Protocol and its parties showed that if there is a will and commitment, humanity can change the environment to its pre-disturbed state to sustain life on this planet.

5

ENERGY PROBLEMS-SOLUTIONS -SUSTAINABILITY OF HUMANITY

Energy Efficiency and Options

Energy is the ability to change or move matter. Energy efficiency means changing or moving the most matter with the least amount of energy. Energy efficiency technologies refer to products or systems designed to use less energy for the same or higher performance than regular products or systems. Negawatt power is a term coined and introduced by Amory Lovins in 1989. He popularized the notion of *negawatts* – the idea of meeting energy needs by increasing efficiency instead of increasing energy production. This technique works by utilizing consumption efficiency to increase available market supply rather than by increasing plant generation capacity. This "virtual generation" method can supply growth of supply by increasing efficiencies of electric energy rather than increasing generation. It has two main uses -

1. The electrical energy available as a result of the application of consumption efficiencies.
2. An arbitrage way of supplying additional electrical energy to consumers without increased generation

capacity by the creation of a market for trading of increased efficiency.

While it is related to and utilizes consumption efficiencies, it differs in scale and market behavior from individual company or consumer level efficiencies. Energy consumers may also reduce energy consumption for a few hours to "generate" negawatts - hypothetical tradeable units of saved energy. By shutting off air conditioners energy can be saved over a short period of time. This reduction in consumption is referred to as a negawatt and can be sold in certain specialized markets. Negawatts represent power saved from one application that is made available to another application. For example, a compact fluorescent light bulb uses about a fourth as much energy as a standard incandescent bulb to put out a similar amount of light. Replacing one 100 watt bulb with one 25 watt compact fluorescent therefore "generates" 75 negawatts of saved energy to use somewhere else. The Negawatt revolution is a market driver conservation mechanism which will create a positive and sustainable solution in the long-term. Energy productivity which measures the output and quality of goods and services per unit of energy input - can come from either reducing the amount of energy required to produce something, or from increasing the quantity or quality of goods and services from the same amount of energy. It can achieve real emission reductions at low cost.

Options to mitigate green house gas emissions are the use of alternative fuels. The alternative fuels commercially available for use for vehicles include the following:

1. Biodiesel
2. Electricity
3. Ethanol
4. Hydrogen
5. Methanol
6. Natural Gas
7. Propane
8. Compressed Natural Gas

Biodiesel

It refers to a non-petroleum-based diesel fuel consisting of short chain alkyl (methyl or ethyl) esters, made by trans-esterification of vegetable oils or animals fats, which can be used alone or blended with conventional petro-diesel in unmodified diesel-engine vehicles. Much of the world uses a system known as the "B' factor to state the amount of bio-diesel in any fuel mix: fuel containing 20% biodiesel is labeled B20, while pure bio-diesel is referred to as B100. Biodiesel can also be used in its pure form (B100), but may require certain engine modifications to avoid maintenance and performance problems. Blending B100 with petro-diesel may be accomplished by mixing in tanks at manufacturing point prior to delivery to tanker truck, by splash mixing in the tanker truck (adding specific percentages of biodiesel and petrodiesel), and by in-line mixing, two components arrive at tanker truck simultaneously. The use of biodiesel provides energy security and emission and safety benefits. Since it is a renewable source and also comes from bioresources, it is an effective and important alternative fuel to reduce greenhouse gas emissions. The use of biodiesel in a conventional diesel engine results in substantial reduction of unburned hydrocarbons, carbon monoxide, and particulate matter. It also decreases the solid carbon fraction of particulate matter since the oxygen in biodiesel enables more complete combustion to carbon dioxide, eliminates the sulfate fraction as there is no sulfur in the fuel, while the soluble or hydrocarbon, fraction stays the same or is increased.

Electricity

Electricity can be used to power electric and plug-in hybrid electric vehicles directly from the power grid. Vehicles that run on electricity produce no tailpipe emissions. The only emissions that can be attributed to electricity are those generated in the production process at the power plant. Electricity is easily accessible for short-range driving. It is considered an alternative fuel since it is used to charge the batteries that power the motors in electric vehicles. Electric cars can also be powered by hydrogen fuel cells instead of batteries. Even though the vehicle itself produces zero pollutants, when emissions from the power generating stations from traditional sources (coal, oil-fired or nuclear) are factored in, battery powered electric cars still produce less than 10 per cent of the emissions of standard internal combustion engine cars.

Ethanol

Ethanol (ethyl alcohol) is a clear, colorless liquid with a characteristic, agreeable odor. Ethanol is also made into ether, ethyltertiary-butyl ether, which has properties of interest for oxygenated gasoline and reformulated fuels. The environmental benefits of ethanol include: 10% ethanol blends reduce carbon monoxide better than any other reformulated gasoline blend. Ethanol is a safe replacement for toxic octane enhancers in gasoline such as benzene, toluene and xylene. Ethyltertiary-butyl ether lowers gasoline volatility and is, thus, particularly effective in reducing volatile organic carbon emissions from automobiles.

Hydrogen

Hydrogen has the potential to revolutionize transportation and, possibly, our entire energy system. The simplest and most abundant element in the universe, hydrogen can be produced from fossil fuels and biomass and even by electrolyzing water. Producing hydrogen with renewable energy and using it in fuel cell vehicles holds the promise of virtually pollution-free transportation and independence from imported petroleum. The following are some ways to produce hydrogen.

Natural gas reforming—"synthesis gas" is created by reacting natural gas with high-temperature steam or by partial oxidation, and then the synthesis gas is reacted with water to produce hydrogen.

- Renewable electrolysis—an electric current generated by renewable energy technologies, such as wind or solar, splits water into hydrogen and oxygen.

- Gasification—Coal or biomass is converted into gaseous components and then into synthesis gas, which is reacted with steam to produce hydrogen.
- Renewable liquid reforming—renewable liquid fuels such as ethanol are reacted with high-temperature steam to produce hydrogen near the point of end-use.
- Nuclear high-temperature electrolysis—heat from a nuclear reactor is used to improve the efficiency of water electrolysis to produce hydrogen.
- High-temperature thermo-chemical water-splitting—high temperatures generated by solar concentrators or nuclear reactors drive chemical reactions that split water to produce hydrogen.
- Photobiological—microbes such as green algae consume water in the presence of sunlight, producing hydrogen as a byproduct.
- Photoelectrochemical—photoelectrochemical systems produce hydrogen from water using special semiconductors and energy from sunlight.

Hydrogen can be produced from diverse domestic resources, with the potential for near-zero greenhouse gas emissions. Once produced, it generates power in fuel cells without exhaust emissions. It holds promise for economic growth in both the stationary and transportation energy sectors.

Methanol

It is methane with one hydrogen molecule replaced by hydroxyl radical. It is also known as wood alcohol which can be made from various biomass resources like wood and coal. Today, nearly all methanol is made from natural gas, because it is cheaper. It can be used as an alternative fuel. Methanol is also very poisonous and very harmful if swallowed. Methanol must not be confused with ethanol. As with gasoline, it is also wise to avoid skin contact with methanol, as it can pass through the skin. The use of methanol has dramatically declined since the early 1990s, and auto makers are no longer manufacturing vehicles that run on it. Methanol fuel cells will greatly reduce carbon dioxide emissions for vehicles and virtually eliminate smog and particulate pollution.

Natural gas

It is a gaseous fossil fuel consisting primarily of methane but including significant quantities of ethane, propane, butane and pentane. Natural gas is an important alternative fuel and is readily available to end users through the utility infrastructure. It can produce significantly fewer harmful emissions than gasoline or diesel when used in natural gas vehicles.

Propane

Liquified petroleum gas (LPG) consists mainly of propane, propylene, butane, and butylene in various mixtures. It is produced as a by-product of natural gas processing and petroleum refining. With propane's simple molecular composition, propane - fueled vehicles emit significantly lower levels of carbon monoxide, hydrocarbons and nitrogen oxides than gasoline - fueled vehicles. The level of air - toxic emissions from propane-fueled vehicles is also low. It has a high energy density, giving propane vehicles good driving range, and propane fueling infrastructure is widespread.

Compressed Natural Gas

It is a substitute for gasoline (petrol), diesel or propane fuel. It is considered to be an environmentally "clean" alternative to those fuels and it is much safer than other motor fuels in the event of a fuel spill: natural gas is lighter than air, so it disperses quickly when leaked or spilled. It is made by compressing natural gas (which is mainly composed of methane), to less than 1% of its volume at standard atmospheric pressure. It is stored and distributed in hard containers, at a normal pressure of 200–220 bar, usually in cylindrical or spherical shapes to maintain equal pressure on the walls of the containers. In response to high fuel prices and environmental concerns, compressed natural gas is starting to be used in light-duty passenger vehicles and pickup trucks, medium-duty delivery trucks, and in transit and school buses. Compressed natural gas vehicles require a greater amount of space for fuel storage than conventional gasoline power vehicles. Since it is a compressed gas, rather than a liquid like gasoline, CNG takes up more space for each GGE (Gallon of Gas Equivalent). Therefore, the tanks used to store the CNG usually take up additional space in the trunk of a car or bed of a pickup truck which runs on CNG. While CNG-powered vehicles are considered to be safer than gasoline-powered vehicles, there are concerns about how best to fight fires involving CNG vehicles.

CNG is often confused with liquified natural gas (LNG). While both are stored forms of natural gas, the key difference is that CNG is in compressed form, while LNG is in liquefied form. CNG has a lower cost of production and storage compared to LNG as it does not require an expensive cooling process and cryogenic tanks. CNG requires a much larger volume to store the same mass of gasoline or petrol and the use of very high pressures (205 to 275 bar).

Several other fuels of importance in view of the present global warming and climate change are currently under development. Many of these fuels are also considered alternative fuels and may have other benefits such as reduced emissions or increased energy security.

1. Biobutanol
2. Biogas
3. Biomass to Liquids (BTL)
4. Coal to Liquids (CTL)
5. Gas to Liquids (GTL)

Biobutanol

Like ethanol, biobutanol is a liquid alcohol that can be produced through processing of domestically grown crops, such as corn and sugar beets, and other biomass, such as fast-growing grasses and agricultural waste products. Biobutanol can be used in gasoline-powered internal combustion engines. The properties of biobutanol make it highly amenable to blending with gasoline. It is also compatible with ethanol blending and can improve the blending of ethanol with gasoline. The energy content of biobutanol is 10 to 20 per cent lower than that of gasoline. Its use reduces greenhouse gas emissions because carbon dioxide captured when the feedstock crops are grown balances carbon dioxide released when biobutanol is burned.

Biogas

Biogas is produced from the anaerobic digestion of organic matter such as animal manure, sewage, and municipal solid waste. After it is processed to required standards of purity, biogas becomes a renewable substitute for natural gas and can be used to fuel natural gas vehicles. Biogas is the gaseous product of the anaerobic digestion (decomposition without oxygen) of organic matter. It is typically made up of 50-80% methane, 20-50% carbon dioxide, and traces of gases such as hydrogen, carbon monoxide, and nitrogen. In contrast, natural gas is typically made up of more than 70% methane, with most of the rest being other hydrocarbons (such as propane and butane) and only small amounts of carbon dioxide and other contaminants. Biogas is sometimes called swamp gas, landfill gas, or digester gas. When its composition is upgraded to a higher standard of purity, it can be called renewable natural gas. Biogas is used for many different applications worldwide. In rural communities, small-scale digesters provide biogas for single-household cooking and lighting. China alone is estimated to have 8–17 million of these systems. Large-scale digesters provide biogas for electricity production, heat and steam, chemical production, and vehicle fuel. The sources for biogas production include sewage sludge, agricultural wastes, industrial wastes, animal by-products and municipal solid wastes. The benefits of biogas are similar to the benefits of natural gas: increasing energy security, paving the way for fuel cell vehicles, and improving public health and the environment through reduced vehicle emissions. The following are additional potential benefits of biogas:

- It is a domestic, renewable resource; using it offsets the use of non-renewable resources such as coal, oil, and fossil fuel-derived natural gas, with corresponding emission reduction and energy security benefits.
- It directly reduces greenhouse gas emissions by preventing methane release into the atmosphere.
- Its production creates jobs and benefits the local economy.
- Anaerobic digestion systems (non-landfill) treat waste naturally, require less land area than aerobic composting, reduce the amount of material that must be landfilled, reduce waste odors, and produce sanitized compost and nutrient-rich liquid fertilizer.

Biomass to Liquids

Biomass to liquids is a term describing processes for converting diverse biomass feedstocks into a range of liquid fuels. These processes are usually distinguished from enzymatic/fermentation processes and processes that use only part of a biomass feedstock, such as those typically used to produce ethanol, biobutanol and biodiesel. Currently, the major biomass-to-liquids production processes are gas-to-liquids processes, involving conversion of biomass into gas and then into liquids, and pyrolysis, involving decomposition of biomass in the absence of oxygen to produce a liquid oil. Biomass-to-liquids processes have the potential to produce a wide range of useful fuels and chemicals. These include transportation fuels such as gasoline, diesel, and ethanol. These fuels are cleaner and more efficient than the solid biomass from which they were derived. Another benefit is that biomass-to-liquids processes can convert types of biomass, such as wood and agricultural residues, that are difficult to handle using other biofuel production processes.

The problems of climate change and the rising cost of oil

have led to a race to develop environmentally friendly bio-fuels such as palm oil or ethanol derived from corn and sugarcane. The European Union has proposed that 10% of the fuel used in transport should come from bio-fuels by 2020 and the emerging global market is expected to be worth billions of dollars a year. But the new fuels have attracted controversy. Using bio-fuels made from corn, sugarcane and soy could have a greater environmental impact than burning fossil fuels. Although the fuels themselves emit fewer greenhouse gases, they all have higher costs in terms of biodiversity loss and destruction of farmland. Regardless of how effective sugarcane is for producing ethanol, its benefits quickly diminish if carbon-rich tropical forests are being razed to make the fields thereby causing vast green house gas emissions.

Food prices will rise in the next 10 years almost twice as sugarcane, maize and oilseed rape are grown to fuel cars and as people in rapidly developing countries adopt meat based diets. The move to agro fuels are expected to marginally lower climate change emissions and reduce oil dependence. In 2007, more than a third of the total U.S. maize crop went to ethanol for fuel, a 48 per cent increase on 2005. Brazil and China grew the crops on nearly 20 million hectares. This area could double in 10 years. But, the switch to growing fuel crops will take land out of food production and increase the price of commodities such as sugar, maize and palm oils. While higher food prices are profitable for the mainly large scale farmers who grow them, they threaten the economies of food importing countries as well as the urban poor. The higher food prices will also mean extra costs for livestock farmers who must buy feed. The price of wheat and some other food is edging record levels after devastating weather in Australia, the running down of grain reserves in the U.S. and drought in Africa. Food price inflation stands at more than 6 per cent a year in some developing countries. Continuing growth will increase the amount of meat reared. Nearly 30 per cent more beef, 50 per cent more pig meat and 25 per cent more poultry are expected to be consumed in developing countries by 2016 with 70 per cent more skimmed milk powder and sugar. The rush to energy crops was encouraging intensive industrial agriculture at the expense of sustainable food production.

Coal to Liquids

Coal to liquids is a term describing processes for converting coal into liquid fuels such as gasoline and diesel. Coal-derived gasoline and diesel could be transported through existing pipelines, dispensed at existing fueling stations, and used to fuel today's gasoline- and diesel-powered vehicles. Producing liquid fuels from coal results in substantial reduction in greenhouse gas emissions; carbon sequestration technologies can mitigate this impact.

Gas to Liquids

Biomass is turned into a gas by heating it in the presence of about one third of the oxygen necessary for complete combustion. The result is a mixture of carbon monoxide and hydrogen called synthesis gas or syngas. Syngas can be burned directly to produce electricity or converted into hydrocarbons (such as gasoline and diesel), alcohols, ethers, or chemical products. In pyrolysis process, biomass is heated in the absence of oxygen to produce liquid pyrolysis oil—sometimes called bio-oil—which can be burned like fuel oil or refined into chemicals and fuels.

Clean Coal

Clean coal is the name attributed to coal chemically washed of minerals and impurities, sometimes gasified, burned and the resulting flue gases treated with steam with the purpose of removing sulfur dioxide, and reburned so as to make the carbon dioxide in the flue gas economically recoverable. The coal industry uses the term "clean coal" to describe technologies designed to enhance both the efficiency and the environmental acceptability of coal extraction, preparation and use with no specific quantitative limits on any emissions, particularly carbon dioxide. The most promising "clean coal" technology involves using the coal to make hydrogen from water, then burying the resultant carbon dioxide by-product and burning the hydrogen. The greatest challenge is bringing the cost of this down sufficiently for "clean coal" to compete with nuclear power on the basis of near-zero emissions for base-load power.

- Coal cleaning by 'washing' has been standard practice in developed countries for some time. It reduces emissions of ash and sulfur dioxide when the coal is burned.
- Electrostatic precipitators and fabric filters can remove 99% of the fly ash from the flue gases - these technologies are in widespread use.
- Flue gas de-sulfurization reduces the output of sulfur dioxide to the atmosphere by up to 97%, the task depending on the level of sulfur in the coal and the extent of the reduction. It is widely used where needed in developed countries.
- Low-NOx burners allow coal-fired plants to reduce nitrogen oxide emissions by up to 40%. Coupled with re-burning techniques NOx can be reduced 70% and selective catalytic reduction can clean up 90% of NOx emissions.
- Increased efficiency of plant - up to 45% thermal efficiency now (and 50% expected in future) means that newer plants create less emission per kWh than older ones.
- Ultra-clean coal from new processing technologies which reduce ash below 0.25% and sulfur to very low levels mean that pulverized coal might be fed directly into gas turbines with combined cycle and burned at high thermal efficiency.
- Gasification, including underground gasification *in situ*, uses steam and oxygen to turn the coal into carbon monoxide and hydrogen.
- Sequestration refers to disposal of liquid carbon

dioxide, once captured, into deep geological strata.

Some of these impose operating costs without concomitant benefit to the operator, though external costs will almost certainly be increasingly factored in through carbon taxes or similar which will change the economics of burning coal.

The burning of coal for power generation was a major source of emissions of green house gases and clean coal technologies could help reduce the pollution. Electricity consumption could be reduced by about 25 per cent with greater energy efficiencies in manufacturing processes, houses and offices and by using better irrigation pumps. There has to be greater use of renewable sources of energy. Mass transportation systems needed to be supported while discouraging individual transportation. The use of bio-fuels for vehicular traffic is to be promoted. Development of railway lines and frequency of trains are an important means to reduce consumption of fossil fuels in transport sector.

Carbon Credits

The Kyoto Protocol has created a mechanism under which countries that have been emitting more carbon and other gases (greenhouse gases include ozone, carbon dioxide, methane, nitrous oxide and even water vapour) have voluntarily decided that they will bring down the level of carbon they are emitting to the levels of early 1990s.

The Kyoto Protocol provides for three mechanisms that enable countries or operators in developed countries to acquire greenhouse gas reduction credits:

- Under Joint Implementation a developed country with relatively high costs of domestic greenhouse reduction would set up a project in another developed country.
- Under the Clean Development Mechanism, a developed country can 'sponsor' a greenhouse gas reduction project in a developing country where the cost of greenhouse gas reduction project activities is

usually much lower, but the atmospheric effect is globally equivalent. The developed country would be given credits for meeting its emission reduction targets, while the developing country would receive the capital investment and clean technology or beneficial change in land use.

- Under International Emissions Trading, countries can trade in the international carbon credit market to cover their shortfall in allowances. Countries with surplus credits can sell them to countries with capped emission commitments under the Kyoto Protocol.

These carbon projects can be created by a national government or by an operator within the country. In reality, most of the transactions are not performed by national governments directly, but by operators who have been set quotas by their country.

For trading purposes, one allowance or CER is considered equivalent to one metric ton of CO_2 emissions. These allowances can be sold privately or in the international market at the prevailing market price. These trade and settle internationally and hence allow allowances to be transferred between countries. Climate exchanges have been established to provide a spot market in allowances, as well as futures and options market to help discover a market price and maintain liquidity. Carbon prices are normally quoted in Euros per ton of carbon dioxide or its equivalent (CO_2e). Other greenhouse gases can also be traded, but are quoted as standard multiples of carbon dioxide with respect to their global warming potential. These features reduce the quota's financial impact on business, while ensuring that the quotas are met at a national and international level.

Carbon credits create a market for reducing greenhouse emissions by giving a monetary value to the cost of polluting the air. Emissions become an internal cost of doing business and are visible on the balance sheet alongside raw materials and other liabilities or assets.

By way of example, consider a business that owns a factory putting out 100,000 tons of greenhouse gas emissions in a year. Its government is an Annex I country that enacts a law to limit the emissions that the business can produce. Then, the factory is given a quota of say 80,000 tons per year. The factory either reduces its emissions to 80,000 tons or is required to purchase carbon credits to offset the excess. After costing up alternatives, the business may decide that it is uneconomical or infeasible to invest in new machinery for that year. Instead it may choose to buy carbon credits on the open market from organizations that have been approved as being able to sell legitimate carbon credits.

- One seller might be a company that will offer to offset emissions through a project in the developing world, such as recovering methane from a swine farm to feed a power station that previously would use fossil fuel. So although the factory continues to emit gases, it would pay another group to reduce the equivalent of 20,000 tons of carbon dioxide emissions from the atmosphere for that year.
- Another seller may have already invested in new low-emission machinery and have a surplus of allowances as a result. The factory could make up for its emissions by buying 20,000 tons of allowances from them. The cost of the seller's new machinery would be subsidized by the sale of allowances. Both the buyer and the seller would submit accounts for their emissions to prove that their allowances were met correctly.

Carbon credits are a part of international emission trading norms. They incentivise companies or countries that emit less carbon. The total annual emissions are capped and the market allocates a monetary value to any shortfall through trading. Businesses can exchange, buy or sell carbon credits in international markets at the prevailing market price.

Carbon credits are a key component of national and

international emission trading schemes that have been implemented to mitigate global warming. They provide a way to reduce greenhouse effect emissions on an industrial scale by capping total annual emissions and letting the market assign a monetary value to any shortfall through trading. Credits can be exchanged between businesses or bought and sold in international markets at the prevailing market price. Credits can be used to finance carbon reduction schemes between trading partners and around the world.

There are also many companies that sell carbon credits to commercial and individual customers who are interested in lowering their carbon footprint on a voluntary basis. These carbon offsetters purchase the credits from an investment fund or a carbon development company that has aggregated the credits from individual projects. The quality of the credits is based in part on the validation process and sophistication of the fund or development company that acted as the sponsor to the carbon project. This is reflected in their price; voluntary units typically have less value than the units sold through the rigorously-validated Clean Development Mechanism.

India and China are likely to emerge as the biggest sellers and Europe is going to be the biggest buyers of carbon credits. India is one of the countries that have 'credits' for emitting less carbon. India and China have surplus credit to offer to countries that have a deficit. India has generated some 30 million carbon credits and has roughly another 140 million to push into the world market. Waste disposal units, plantation companies, chemical plants and municipal corporations can sell the carbon credits and make money. More than 112 Indian companies, including Hindustan Lever Ltd and Tata Steel, are set to trade in carbon credits. These companies are ready with clean technologies to bring down the emission levels of greenhouse gases and sell certified emission reductions (CERs) to developed countries. According to World Bank estimates, India is expected to rake in $100 million annually by trading in carbon credits and Indian companies are expected to corner at least 10 per cent of the global market in the initial years.

India is the world's sixth largest emitter of carbon dioxide with its present share in global emissions estimated at 6 per cent.

- India is expected to rake in $100 million annually by trading in carbon credits.
- Indian firms are expected to corner 10% of the global market in the initial years.
- They are likely to generate $8.5 billion at the going rate of $10/ton of certified emission reductions.

Globally, greenhouse gas emissions are expected to come down by 2.5 billion tons by 2012. According to industry estimates, Indian companies are expected to generate at least $8.5 billion at the going rate of $10 per ton of CER.

Urban Land Use Changes

Changes in land use in urban areas are important to counter the effects of global warming. Parks are purely ornamental and recreational. They are also water intensive and cater to fixed strata of the city populace. Urban agriculture is the opposite of city beautification projects such as parks. It means horticulture, wetland development, fish farms and domestic animal farms. Apart from several economic benefits, it helps to sequester carbon dioxide from the atmosphere. It helps with poverty alleviation and food security besides being recreational and educational.

The percentage of urban poor in Asian cities averages about 20 to 40 per cent. In Indian cities, the slum population can be as high as 55 per cent. These large numbers of poor usually pay high prices for low quality food. According to FAO estimates, low income urban households spend 60 to 80 per cent of their budget on food which is 30 per cent more than rural households. The World Food Summit in 1996 has highlighted the need to increase access to food in the wake of increasing urbanization. Food security is not only about availability but also concerns accessibility. High food prices mean low accessibility.

The problems are further compounded by the way food distribution in cities is fast changing. Increasingly, city planners are doing away with hawkers and other informal food distribution points. Administrators typically consider hawkers and street food vendors a liability - as something that takes away from their cities image. Instead, cities increasingly promote supermarkets and other large-scale organized shops. These cater to higher income groups and rely on food processing and packing. Processed foods increase the ecological footprint of a city, as goods have to be transported from long distances.

The Food and Agriculture Organization estimates that Delhi will require an additional 1, 96,500 trucks of 10 tons capacity by 2010 to supply food for its population while Mumbai will need an additional 3, 13,400 trucks of supply a year. This will have serious implications for traffic and roads.

Cities that have seriously considered the issue of urban poverty environment and food security have made plans to allow for more agriculture within their urban and peri-urban areas. Bangkok has 60 per cent of its metropolitan area as agriculture land, as has Madrid. Beira in Mozambique has a high percentage of about 88 per cent of its green spaces used for family agriculture. Ottawa has 5,000 hectares of agriculture land within city limits. In the United States, cities such as Minneapolis, Austin, and Boston have provisions in the zoning laws for community gardens. The famous marshes of Xochimilco, located on the outskirts of Mexico City are fed by treated wastewater from the city. This water is used for irrigating flowers and vegetables and also recharging the aquifers. With a careful selection of plant species, the agricultural yield can be increased. In certain cases, the yield has been increased. In certain cases, the yield has been as high as 50 kg of fresh produce per square meter a year.

Vision for Indian cities is not encouraging. The proposed Delhi Master Plan 2021 has allocated only 15 to 20 per cent under green and recreational areas. Much of it is meant for parks and recreational areas. In Bangalore, the 742 sq km of green belt around the city is reduced to 248 sq km in the proposed master plan. The emphasis is clearly on land for buildings. Therefore, urban agriculture and urban forestry need to be promoted in urban and sub-urban lands in order to provide livelihood to poor and the needy and at the same time to sequester carbon dioxide from vehicular and industrial chimneys. These steps will reduce the burden on rural farmers who supply all the food to urbanites and restore the soil cover, vegetation cover, water table and climate as well in urban areas.

Change the Diet of Ruminants to Cut Methane Emissions

Agriculture accounts for about 14% of the greenhouse gases into the atmosphere. Cows, goats, sheep and several other animals belong to a class of animals called ruminants. Ruminants have four stomachs and digest their food in their stomachs instead of in their intestines, as humans do. Ruminants eat food, regurgitate it as cud and eat it again. The stomachs are filled with bacteria that aid in digestion, but also produce methane. The world's 1.5 billion cows and billions of other grazing animals emit dozens of polluting gases, including lots of methane. Two-thirds of all ammonia comes from cows. Cows emit a massive amount of methane through belching, with a lesser amount through flatulence. A single cow can produce between 100 and 200 liters of methane a day. In any case, that is a lot of methane, an amount comparable to the pollution produced by a car in a day.

Initially, grazing areas were filled with a variety of grasses and flowers that grew naturally, offering a diverse diet for cows and other ruminants. However, in order to improve the efficiency of feeding livestock, many of these pastures became reseeded with perennial grasses. With the aid of artificial fertilizers, perennial grasses grow quickly and in huge quantities. The downside is that it lacks the nutritious content of other grasses and prevents more nutritious plants from growing. It is the "fast food" of grasses. This simple diet allows many cows and other ruminants to be fed, but it inhibits digestion. A perennial grass diet also results in a significant number of weak and infertile ruminants, which have to be killed at a young age. This is where the methane comes in. The difficult-to-digest grass ferments in the ruminants' stomachs, where it interacts with microbes and produces gas. These ruminants are methane-making machines, with their inefficient digestion producing hundreds of liters of the greenhouse gas every day. Burping cows and sheep produce more methane. Farmed ruminant animals are thought to be responsible for up to a quarter of "man-made" methane emissions worldwide though, contrary to common belief, most gas emerges from their front, not rear, ends. Recent researches indicate that the diet of farmed animals can be changed to make them produce less methane. Farmers could help tackle climate change by growing grass varieties bred to have high sugar levels for their animals to eat. The altered diet changes the way that bacteria in the stomachs of the animals break down plant material into waste gas.

A recent study compared three types of naturally grown pastures to ryegrass pasture grown with chemical fertilizers. Lambs were fed on each type of pasture. The meat from lambs fed on natural pastures had less saturated fat, more omega-3fatty acids, more vitamin E and higher levels of conjugated linoleic acid, a "good fat" that is believed to fight cancer. The meat from these lambs was considered very high quality and scored well in flavor tests. Because of concerns about ruminant diets, many researchers are investigating ways to alter what livestock eat and to mix the best of old cow pastures – diverse, naturally growing, nutrient rich grasses and plants – with the best of the new – fast-growing and resistant to invasive species. One possibility is to increase the ability of beneficial, nutrient-rich plants and flowers to grow alongside the fast-growing grasses commonly used in pastures. Another branch of research focuses on plants that are high in tannins, which are believed to lower methane levels in ruminants and to boost milk production – although excessively high level of tannins are harmful to a ruminant's growth. Planting legumes and genetically engineered plants to trap airborne nitrogen will also improve nitrogen levels in the soil, which is important for rich soil and healthy plants. Believers in naturally grown, mixed-species pastures say that the use of them will reduce greenhouse gases, improve animal health and meat quality and reduce the use of artificial fertilizers. Efforts like methane-reducing pills or the addition of garlic may just be stopgap measures that fail to address some of the core problems of livestock, namely ground and air pollution, cutting down of forests, the production of weak animals that later have to be culled and the use of artificial fertilizers and steroids. Another possibility exists in trapping the methane gas and using it as energy or selling it back to the electrical grid. Some farmers already extract methane from livestock waste, but that does not solve the bigger problem of belched methane. Harnessing that methane would mean trapping it in the air, perhaps by housing cattle indoors or outfitting them with special muzzles that may inhibit eating. Attempts are also being made to develop plant-based pills combined with a special diet and strict feeding times to reduce the methane produced by ruminants by converting it to glucose. By changing the feed material which is easily digestible, methane emissions from farm animals could be reduced by 50 per cent.

Rooftop Gardens

Rooftop gardens involve covering the rooftop with a layer of soil on top of a waterproof membrane and cultivating a garden with small plants. Such gardens may even be used to grow vegetables. These gardens are highly beneficial. They help in sequestering carbon dioxide and reduce rainwater runoff. They insulate the buildings, reduce the need for air-conditioning and provide breathing space in cities.

Green roofs serve as sponges, to absorb run-off water, and lessen the impact of severe storms that otherwise may cause damage to the local environs, hasten sewers to overflow and hence pollute the environment. When properly designed, green additions are capable of cooling structures and the surrounding neighborhoods, by the process of evapo-transpiration; the key way plants absorb water through their root system and evaporate it through their leaves. Not only do green roofs lower the temperatures in the summer, but they also insulate in the winter. On average, extensive green roofs provide twenty-five per cent more insulation than a regular roof, while heat loss due to wind can be reduced by fifty per cent. In addition, the gardens filter pollutants from the air, and offer opportunities for food production, as well. All the wasted space on city rooftops can be like small oases. They can help with storm water management and the urban heat island effect, or serve as places to garden. As green roofs spread, the overall effect may help mitigate global warming. By reducing the urban heat island effect, the carbon footprint of energy used for air conditioning, for instance, will be lessened. In its need for variety and acceptance of randomness, a flourishing natural ecosystem is more like a city than a plantation. Perhaps it will be the city that reawakens our understanding and appreciation of nature, in all its teeming, unpredictable complexity. The promotion of green rooftops concept is an effective and less expensive way to sequester carbon dioxide and other pollutants from the atmosphere and reverse global warming phenomenon to some extent.

The water-permeable pavement system is a state-of-the-art technology and developed based on the idea from the traditional Japanese cooling method of sprinkling water on streets. In this system, rainwater seeps through the road surface, accumulates under the pavement and percolates back through the pavement by capillary action, cooling down the street by releasing heat through water vaporization. During daytime in summer, the surface temperature of water-permeable pavements is as much as 25 degrees Celsius lower than asphalt surfaces, and reflected heat is reduced by 20 to 40 per cent. The infiltration of water through the pavement mitigates both the heat and the glare.

Planting Trees to Combat Climate Change

Planting tree is hugely beneficial to the world in the face of accelerating climate change and is a key weapon against global warming. There is now a strong international scientific consensus that human activity is causing global warming. A substantial reduction in the planet's forest cover over recent centuries is a major contributor to this climatic change. As trees grow they absorb carbon dioxide, the main 'greenhouse gas' responsible for global warming, thereby reducing the concentration of this gas in the atmosphere. Forests are referred to as 'carbon sinks' for this function of carbon dioxide absorption and storage. Planting trees to bolster carbon sink area, an example of what is termed 'carbon sequestration', helps offset the loss of native forests and fights global warming. At a local level, tree-planting on deforested lands creates further environmental benefits. Forests play a vital role in regulating water supplies, helping to minimize both water shortages in times of drought and damaging floods in heavy rains. Trees also reduce soil erosion, thereby conserving soil quality upstream and water quality downstream. Forests also provide habitats for a wide array of plant and animal species, a number of which are threatened with extinction by deforestation.

Socio-economic benefits of reforestation projects include direct employment, infrastructure development, skills-transfer and the creation of markets for related products and services. Certain plantation forest models may also provide local communities with additional products such as fuel-wood, fruit, nuts and herbs, and opportunities for agricultural activities (inter-cropping, livestock grazing) within the forest area.

Finally, sustainable forestry can provide additional climate change and local environmental benefits in the longer term. Once harvested, the wood from the trees may be used either as a source of renewable energy, thereby reducing the use of fossil fuels, or for construction materials or furniture-making, thereby reducing deforestation elsewhere.

Carbon absorption rates in trees depend on the species and the location. Fast-growing species in tropical climates can sequester carbon dioxide many times faster than the average European forest, and plantation projects aimed at earning carbon credits are typically designed to maximize the sequestration potential. Over a 20-year lifecycle, the right species in the right conditions can absorb over 40,000 tons of CO_2 per square kilometer. A plantation of 100 square kilometers can absorb 4 million tons of CO_2 over 20 years. That is equivalent to taking 50,000 cars or more off the road during that time (based on annual emissions of 3 to 4 tons for the average car and its usage). Even when allowing for other factors that reduce the net carbon effect of tree planting, such as the greenhouse-gas emissions from vehicles and machinery used during planting and maintenance, a sizable plantation makes a very significant positive contribution to the greenhouse problem.

Tree plantations are bad for biodiversity and the local environment. Large mono-culture plantations of exotic species do not benefit native plants and animals, and native forests are sometimes cut down to make way for them. This is true of some commercial timber plantation operations, but not of tree-planting overall, and particularly not of well-regulated carbon-financed projects.

Species selected must be appropriate for the sites, and many schemes require a certain proportion of trees planted to be indigenous species or a certain proportion of the land to be devoted to native forest regeneration. And since offset projects are generally located on land that is already degraded, they still lead to an overall increase in ecosystem health and biodiversity. It is finally worth noting that it is somewhat unfair to judge commercial timber plantations against the benchmark of natural forest. This imposes a higher standard than that faced by commercial growers of other non-indigenous monoculture crops, such as wheat or maize, and ignores the additional environmental benefits that plantation forests do provide.

In tropical climates, the trees generally replace dense ground vegetation and thus have little impact on solar energy absorption. The trees grow fast and absorb high volumes of CO_2, and higher evaporation levels add a further cooling effect. While forest plantation activity goes on at most latitudes around the globe, carbon projects tend to be located in the tropics, where faster growth rates maximize the rate of carbon absorption and hence the volume of carbon credits generated.

All tree-planting projects, especially those in the developing world, should ensure that wages are fair, that land rights are respected, and that the views and needs of local communities are taken into consideration in project design and management. The socio-economic benefits of projects should far outweigh any genuinely unavoidable negative impacts. The Clean Development Mechanism and the emerging accreditation schemes for voluntary carbon projects include social as well as environmental criteria, and require detailed initial analysis and ongoing monitoring to ensure that these are met.

Some ecologists suggest that planting trees to combat climate change is a waste of time. They report that most forests do not have any overall effect on global temperature while those farthest from the equator could actually be making global warming worse. To plant forests to mitigate climate change outside of the tropics is a waste of time. The carbon dioxide used by trees for photosynthesis helps cool the earth by reducing greenhouse gases in the atmosphere. But forests also trap heat from the sunlight they absorb. Outside a thin band around the equator, forests trap more heat than they help to get rid of by reducing carbon dioxide. In simulation studies, linking carbon dioxide and the heat absorbing effect of trees, these ecologists found forests had different effects on global warming depending on latitude. North of 20 degrees latitude, forests had a direct warming influence that more or less counterbalanced the cooling effect of carbon removal from the atmosphere. Past 50 degrees, forests warmed the Earth by an average of 0.8°C, but in the tropics, forests helped cool the planet by an average of 0.7°C. The forest canopies are relatively dark and absorb most of the sun's rays heat falling on them. Grassland or snowfields however reflect more sun, keeping temperatures lower. Planting trees above 50° latitude such as in Siberia could cover tundras normally blanketed in heat reflecting snow. In the tropical regions, through water evaporating from trees increase cloudiness which helps to keep the planet cool. Therefore, planting trees was diversion letting consumers pollute more. It would be better to transform the way energy was derived and used for instance through investment in renewable and carbon free electricity generation.

Iron is essential for plant growth. Most of it reaches the oceans through winds carrying eroded, iron-rich soils from dry areas on land. But changes in climate and vegetation since the end of the last glacial period are believed to have diminished the iron supply to the ocean - reducing the growth of plankton, which naturally absorbs carbon dioxide from the atmosphere. In certain regions, including the equatorial and north Pacific and the entire Southern Ocean, a simple iron addition does cause phytoplankton to grow rapidly. But tiny zooplanktons, known as "grazers," eat much of the bloom almost as soon as it starts. This begins a chain of recycling that ensues from the sea surface to the seafloor as grazers, krill, fish, whales, and decomposers feed upon each other. Much of the immense carbon prize won by the iron addition quickly leaks back into the atmosphere as carbon dioxide gas. What is critical for the effectiveness of iron fertilization schemes is the amount of organic carbon that actually sinks from the surface and is sequestered in the depths. Only a small percentage of carbon—in the form of dead cells and fecal pellets—falls to the seafloor and stays there, unused, for millennia. A higher percentage (between 20 and 50 per cent) will at least reach middle-depth waters, where the carbon will remain in underwater currents for decades. Beyond the inefficiency of carbon sequestration, iron fertilization would likely cause other changes "downstream" of the ocean patches where iron was added. The huge green phytoplankton blooms would take up not just iron but other nutrients, too—nitrate, phosphate, and silica—essentially depleting nearby waters of the building blocks needed for plankton growth. Large-scale iron fertilization, in altering the base of the food chain, might lead to undesirable changes in fish stocks and whale populations. Increased decomposition of sinking organic matter could deprive deep waters of oxygen or produce other greenhouse gases more potent than carbon dioxide, such as nitrous oxide and methane. The plankton-choked surface waters could block sunlight needed by deeper corals, or warm the surface layer and change circulation patterns. On the other hand, more plankton might produce more of a chemical called dimethylsulfide, which can drift into the atmosphere and encourage cloud formation, thus cooling the atmosphere and helping to counteract greenhouse warming. And others argue that increased plankton supplies might enhance fish stocks. Iron fertilization companies would earn profits by measuring how much carbon they sequester and then selling the equivalent to companies (or people) that either wish to or are required to offset their emissions. Any plan to sell sequestered carbon requires a reliable accounting, and this promises to be difficult in the ocean.

Iron fertilization appears to have the potential to significantly alter plankton blooms, marine ecosystems, and perhaps even the planet. Therefore, understanding the role which iron plays is essential to understanding the biogeochemistry of our oceans and world. Some hope that this has the potential to mitigate global warming while others point out that such large-scale geo-engineering would have many unknown repercussions. While the debate about global warming and what should be done to minimize its effects continues, we now realize that the oceans have a potentially significant part in that drama. Furthermore, iron plays a definitive role in unlocking some of the secrets the oceans once held.

Dealing with the Collapsing Ecosystems

The biodiversity loss was a result of resources being

consumed faster than the planet could replace them. Populations of species in terrestrial, marine and freshwater ecosystems have declined by more than 30% since 1970. In the tropics the declines are even more dramatic, as natural resources are being intensively exploited for human use. If the world's population adopts the lifestyle of USA or UK lifestyle, three planets would be needed to support their needs. It is time to make some vital choices to enable people to enjoy a one planet lifestyle.

Five ways to cut the global ecological deficit are: increases in population can be slowed down and turned into decreases by people having smaller families. The affluent can cut consumption and still improve their quality of life. The resources used in the production of goods can be greatly reduced. And better management can reclaim land while rigorous protection of soils, fisheries and forests will help the planet produce more resources.

Dealing with Coastal Areas

Country-scale maps depicting which areas will require shore protection - dikes, bulkheads, beach nourishment and which areas will be allowed to adapt naturally should be developed. A careful analysis should be done regarding the environmental consequences of shore protection. Shore protection techniques that do not destroy all habitats should be promoted. Measures should be taken for identifying land use to ensure that wetlands migrate as sea level rises in some areas. State and local governments should be engaged in defining responses to sea level rise. Improving early warning systems and flood hazard mapping for storms should be improved. Water supplies should be protected from contamination by salt-water.

The impact of sea level rise on Indian coastline could be significant which is already under stress due to unsustainable human activities. To devise appropriate and effective response strategies, it is essential to make vulnerability assessment studies at regional and local levels. For this, all the factors related to eustatic sea level rise, local hydrological, meteorological and oceanographic changes, local compaction and tectonic movements need to be considered. There is an urgent need to make a systematic inventory on tidal gauge measurements, mud flats and drifts, estuarine activities and other socio-economic activities which are being carried out in coastal zones. Efforts should be initiated to protect the coastal zones form erosion, salt-water intrusion by judiciously choosing the hard (e.g., seawall, revetments, groins, etc.) or soft structures (e.g., regeneration of natural protective systems like mangroves, coral reefs or replenishment of sandy beaches). Coastal zone management plan should also include research and development activities for cost effective methods for the protection of coastal lands. Rules and regulations must be framed and enforced to have a control over the developmental activities and to put restrictions on seaward extrusion. Such tasks should produce a mix of products and services from the available coastal resources, which should enhance India's coastal resilience and facilitate adaptation.

Dealing with Agriculture and Forestry

Farming practices that withstand climate variability - the use of drought-resistant crop varieties and more efficient use of water resources are required to deal with agriculture in view of the climate change resulting from global warming. Certain measures like monitoring world agricultural production for planning and early warning purposes, using satellite technology and agro-metereological tools, establishing effective early warning systems for animal and plant diseases, and monitoring environmental conditions and climatic changes are essential to deal with agriculture. Adaptation measures facilitated by concerted international action to mobilize financing globally and long-term strategic planning in developing countries are urgently needed to reduce the adverse effects of climate change. There are two levels of adaptations: Level 1 adaptations include shifts in planting date that do not imply major changes in crop calendar, additional application of irrigation water to crops already under irrigation, and changes in crop variety to currently available varieties more adapted to the altered climate. The timing of planting dates to adapt to changing growing conditions should be altered. Cropping mix and forest species that are better suited to the changing climatic conditions should be altered. New plant species and crops that are more tolerant to changed climate conditions should be bred. Fire suppression practices in the event of increased fire risk due to temperature increases should be promoted. Insect outbreaks should be controlled. Level 2 adaptations aim at more substantial change to agricultural systems, possibly requiring resources beyond the farmers' means including investment in regional and national agricultural infrastructure and policy changes at the regional and national level. The Level 2 adaptation may significantly reduce the effect of climate change on agriculture and the number of people at risk of hunger. However, adaptation in developing countries, although it does reduce the negative effect of global warming, does not completely eliminate the potential increase in hunger.

Forestry is the science and practice of studying and managing forests and tree plantations and related natural resources. Silviculture, a related science, involves the growing and tending of trees and forests. Modern forestry concerns itself with assisting forests to provide timber as raw material for wood products; wildlife habitat, natural water quality regulation, recreation, landscape and community protection, employment, aesthetically appealing landscapes, biodiversity management, watershed management and a sink for atmospheric carbon dioxide. Forestry involving multiple species of economic and ecological importance is an important remedy for sequestering carbon dioxide and thereby regulating global warming. Long-lived woody plants store carbon in wood and other tissues until they die and decompose from which the carbon in their wood may be released to the atmosphere as carbon dioxide, carbon monoxide or methane or it may be incorporated into the soil as organic matter. Forests with net growth are capable of net absorption of carbon dioxide whereas mature forests with little growth hold carbon stocks but are unable to absorb additional carbon dioxide. Forestry management based on harvesting at the optimal rotation, efficient conversion of the wood into durable products and appropriate regeneration would maximize sequestering of carbon dioxide. Increasing the production of industrial wood products produced from wood obtained on a sustainable basis from managed natural forest, and especially from plantations, would be the most efficient means of using forests to sequester carbon dioxide and thus reversing the global warming phenomenon.

Dealing with Non-CO_2 emissions

Greenhouse gases other than carbon dioxide (non-CO_2 gases) play an important role in efforts to address global climate change due to their high global warming potentials and the availability of cost-effective emission reduction opportunities. Non-CO_2 greenhouse gases are a significant contributor to climate change. Approximately 30% of the human-induced greenhouse effect can be attributed to the non-CO_2 greenhouse gases. Reduction of non-carbon dioxide emissions is to be dealt with by recovering methane emissions from waste management and enhancing carbon storage in sinks. Carbon sinks are natural or man-made processes that remove greenhouse gases from the atmosphere. Carbon storage options include protecting and enhancing storage in natural systems, such as forests and soils; and capturing carbon dioxide produced during energy production and consumption for long-term storage underground or in oceans. The development and widespread use of new technologies plays a major role in climate change mitigation. Just as old technologies, such as coal-fired power stations and internal combustion engines, have contributed to an increase in emissions, so new and more efficient technologies can reduce emissions. Available technologies, such as wind turbines, hybrid engines, and carbon dioxide storage, offer some opportunities to reduce the emissions accumulating in the atmosphere.

Climate Change versus Humanity

For the first time in the history of the world, a very small group with some scientific knowledge can destroy a large chunk of the world's population (millions, if not billions of people) before anyone could stop them. The world is drifting towards a "tipping point" that could lock the world's poorest countries and their poorest citizens in a downward spiral, leaving hundreds of millions facing malnutrition, water scarcity, ecological threats, and a loss of livelihoods. Ultimately, climate change is a threat to humanity as a whole. But it is the poor, a constituency with no responsibility for the ecological debt we are running up, who face the immediate and most severe human costs. For millions of people, these are events that offer a one-way ticket to poverty and long-run cycles of disadvantage. Apart from threatening lives and inflicting suffering, they wipe out assets, lead to malnutrition, and result in children being withdrawn from school. The failure to tackle climate change could leave future generations facing ecological catastrophe.

On mitigation of global warming, the developed countries should demonstrate leadership by cutting greenhouse gas emissions by at least 80% of 1990 levels by 2050. What is required to mitigate the threats arising out of climate change is following a mix of carbon taxation, more stringent cap-and-trade programs, energy regulation, and international cooperation on financing for low-carbon technology transfer.

Human beings are exposed to climate change through changing weather patterns and indirectly through changes in water, air, food quality and quantity, ecosystems, agriculture, and economy. At this early stage, the effects are small but are projected to progressively increase in all countries and regions. The following are some suggested measures to handle the effects of climate change:

- Many diseases and health problems that may be exacerbated by climate change can be effectively prevented with adequate financial and human public health resources, including training, surveillance and emergency response, and prevention and control programs.

- Urban tree planting to moderate temperature increases.
- Weather advisories to alert the public about dangerous heat conditions.
- Grain storage, emergency feeding stations.
- Adjusting clothing and activity levels, increasing fluid intake.

Heat-related deaths can be prevented by emergency measures to move vulnerable people to air-conditioned buildings, and by reducing the emissions of photochemical oxidants which cause ground-level ozone. Many of the impacts of climate change on health could be avoided through the maintenance of strong public health programs to monitor, quarantine, and treat the spread of infectious diseases and respond to other health emergencies as they occur.

Adaptation to environmental change is not a new concept. Human societies have shown throughout history a strong capacity for adapting to different climates and environmental changes. For example, farmers, foresters, civil engineers, and their supporting institutions have been forced to adapt to numerous challenges to overcome adversity or to remove important impediments to sustained productivity. Human-induced climate change represents a new challenge, and may require adaptation approaches to changes that are potentially larger and faster than past experiences with recorded natural climatic variability.

Even if all carbon dioxide emissions stopped at this moment, the amount of carbon dioxide already emitted into the atmosphere will result in an enhanced greenhouse effect for the next 50 years. Thus, people will need to adapt to the effects of climate change. Adaptation is "any action that seeks to reduce the negative effects or to capitalize on the positive effects, of climate change". Adaptive actions may be either anticipatory or reactive in nature. An example of an anticipatory adjustment is the development of heat- and drought-tolerant crop varieties.

Doomsday Seed Vault to Save Humanity

The Svalbard Global Seed Vault is located on the Norwegian island of Spitsbergen near the town of Longyearbyen in the remote Arctic Svalbard archipelago. The island is about 1,120 kilometers from the North Pole. The vault's location deep inside a mountain in the frozen north ensures the seeds can be stored safely no matter what happens outside, providing a Noah's Ark of food crops in the event of a global catastrophe. The Seed Vault is managed under terms spelled out in a tripartite agreement between the Norwegian government, the Global Crop Diversity Trust (GCDT) and the Nordic Genetic Resource Center (previously named the Nordic Gene bank, a cooperative effort of the Nordic countries under the Nordic Council of Ministers). The GCDT has played a key role in the planning of the Seed Vault and is coordinating shipments of seed samples to the Vault in conjunction with the Nordic Genetic Resource Center. The Trust will provide most of the annual operating costs for the facility, and has set aside endowment funds to do so, while the Norwegian government will finance upkeep of the structure itself. The Bill & Melinda Gates Foundation has provided approximately $750,000 to assist developing countries and international agricultural research centers to package and ship seeds to the Seed Vault. An International Advisory Council is being established to provide guidance and advice.

The seed bank was constructed 120 meters inside a sandstone mountain at Svalbard on Spitsbergen Island. The bank employs a number of robust security systems. Seeds are packaged in special four-ply packets and heat sealed to exclude moisture. Spitsbergen was considered ideal due to its lack of tectonic activity and its permafrost which will aid preservation. The location 130 meters above sea level will ensure that the site remains dry even if the icecaps melt. Locally mined coal provides power for refrigeration units that further cool the seeds to the internationally-recommended standard -20 to -30°C. Even if the equipment fails, at least several weeks will elapse before the temperature rises to the -3 °C of the surrounding sandstone bedrock.

The Svalbard Global Seed Vault opened officially on February 26, 2008. Approximately 1.5 million distinct seed samples of agricultural crops are thought to exist. The variety and volume of seeds stored will depend on the number of countries participating. The first seeds arrived in January 2008. It is aimed at safeguarding biodiversity in the face of climate change, wars and other natural and man-made disasters; it can hold up to 4.5 million batches or twice the number of crop varieties believed to exist in the world today. There are currently more than 2, 00,000 different varieties of rice and wheat in the world but this diversity is rapidly disappearing due to pests and diseases, climate change and human activities. Biodiversity is essential because it enables crops to adapt to new conditions, resist diseases, increase their nutritional value and become less dependent on water. The vault is an insurance policy for mankind. There is a need for gene banks around the world that can directly supply plant breeders and researchers with the diversity they need.

The Svalbard Global Seed Vault's mission is to provide a safety net against accidental loss of diversity in traditional gene banks. It can be used in the event of a major regional or global catastrophe; it will certainly be more frequently accessed when gene banks lose samples due to mismanagement, accident, equipment failures, funding cuts and natural disasters. Such events occur with some regularity. In recent years, some national gene banks have also been destroyed by war and civil strife. There are some 1,400 crop diversity collections around the world, but many are in politically unstable or environmentally threatened nations.

The diversity stored in seed gene banks is the raw material for plant breeding and for a great deal of basic biological research. Several hundred thousand samples are distributed annually for such purposes. The United Nation's Food Agriculture Organization lists some 1400 seed banks around the world, the largest being held by the United States Government. Other large banks are held by China, Russia, Japan, India, South Korea, Germany and Canada in descending order of size.

Dubbed the "Doomsday Vault," or the "Noah's Ark" as in the Bible, the seed bank on a remote island near the Arctic Ocean is considered the ultimate safety net for the world's seed collections, protecting them from a wide range of threats including war, natural disasters, and lack of funding or simply poor agricultural management. The seed vault is the perfect place for keeping seeds safe for centuries. At these temperatures, seeds for important crops like wheat, barley and peas can last for up to 10,000 years.

Today is a bleak day for the environment, the day of the year when mankind over-exploits the world's resources - the day when we start living beyond our ecological means. Evidence is mounting that rapid population growth and rising living standards among the Earth's six billion inhabitants are putting an intolerable strain on nature. Man is quickly using the global resources of farming land, forests, fish, air and energy.

The speed at which mankind has used the Earth's resources over the past 20 years has put "humanity's very survival" at risk. The environmental audit, for the United Nations, found that each person in the world now requires a third more land to supply his or her needs than the Earth can supply. The United Nations Organization insists on the right of every human being alive today and to be born in the future to have access to drinking water, to a sufficient and balanced diet, to basic health care, to education and to democracy. This is the minimal base of what is called sustainable development, that is to say the well-being of present generations without compromising that of the generations to come. Today, out of a population of more than 6 billion, 1.2 billion earn less than one dollar a day and 3 billion less than two dollars! Many among them do not benefit from the totality of the rights insisted on by the United Nations.

Meanwhile, in the rich countries, everything around them indicates "well-being is the power to continually buy all sorts of new things". To supply this hyper consumer society with goods and merchandise, it is an inescapable fact that consumption of energy must remain at a very high level and always be on the increase. At current rates, depletion of the greater part of our fossil energy reserves is foreseeable before the end of the 21st century. Certain forecasters think that before the people in rich countries find themselves trapped by the physical reality of limits on resources and therefore enter into a period of compulsory recession, the rich countries could give thought to the possibility of committing themselves to a society with negative growth. This would imply a realization that human happiness can be something other than the plentiful supply of consumer goods whilst preserving growth for those who have not yet obtained the fundamental rights of human beings. But a society of negative growth remains to be defined in its entirety and would necessitate upheavals that we are perhaps not yet ready to accept. Others are not convinced of the imminence of the difficulties and see no reason to question the economic orientations of our world. For them, growth continues to remain the best means of fighting poverty and unemployment. They think that scientific and technical progress in the future will solve the problems before fossil energy sources are depleted. The near future, probably the next 30 years, will decide in favour of one or the other interpretation. In the meantime, it is up to each of us to think about the question, to make our own decisions and, perhaps, to put the fruits of our reflection to work in our daily life, whether it be in one direction or another. First and foremost, it is essential to understand and to analyze the data relevant to the problem, the risks and the hopes. It is an unavoidable personal challenge to anyone who is interested in his or her future. By making use of our knowledge and finding time for reflection, we will be better equipped to respond to the challenges and issues of life. Humanity is living off its ecological credit card and can only do this by liquidating the planet's natural resources. Creation of awareness and conducting education camps for good practices for changes in resources use and lifestyle is inevitable. Some practices are given below for reversing the global warming effects:

1. Use of alternative use of energy -- solar lighting, cooking, water heating, different electrical materials
2. Use of battery operated van
3. Use of biogas for kitchen
4. Promote tree plantation
5. Substitute process material with natural materials
6. Discontinue use of chlorofluorocarbon releasing materials
7. Reduce in greenhouse gas management by good practices like soil/waste management
8. Reduce consumption by using energy efficient application
9. Discourage use of plastics

Therefore, the changes in life styles coupled with reduction, reuse and recycle ways of using natural resources are imperative for the sustainability of the present populations, future populations and the humanity together with biological diversity to go on for ever to perpetuate life on this planet - the EARTH.

"The Earth is like a bed; if you contaminate it, one day you will suffocate and die in your bed"

References

Adam, David. 2007. Move to cut methane emissions by changing cows' diet." The Guardian. July 10th Issue.

Albritton, D. L., Meira Filho, L.G. 2001. Technical Summary. Working Group 1, Intergovernmental Panel on Climate Change. Cambridge University Press, Cambridge.

Aldy, J. E., Scott Barrett and Robert N. Stavins. 2003. Thirteen Plus One: A Comparison of Alternative Climate Policy Architectures. Climate Policy 3 (4):373-397.

Aron, J. L., Corvalan, C. F. and Philippeaux, H. 2003. Climate Variability and Change and Their Health Effects in the Caribbean: Information for Adaptation Planning in the Health Sector. Conference 21-22 May, World Health Organization, Geneva.

Barrett, S. 2003. Environment and Statecraft. Oxford University Press, New York.

Barry, J., Baxter, C. Sagarin, R. and Gilman, S. 1995. Climate related, long-term faunal changes in a Californian rocky inter-tidal community. Science 267: 672-675.

Beniston, M. and Tol, R. S. J. 1998. Europe. In: The Regional Impacts of Climate Change: An Assessment of Vulnerability. pp.149-185, (Eds. R. T. Watson, M. C. Zinyowera and R.H. Moss), Cambridge University Press, Cambridge, UK.

Boyd, P. W. and Law, C. S. 2001. The Southern Ocean Iron Release Experiment (SOIREE) – introduction and summary. Deep-Sea Research (Part II, Topical Studies in Oceanography) 48: 2425-2438.

Brown, B. E.1990. Damage and recovery of coral reefs affected by El Nino related seawater warming in the Thousand Islands, Indonesia. Coral Reefs 8:163-170.

Buesseler, K. O. and Boyd, P.W. 2003. Will Ocean Fertilization Work? Science 300:67-68.

Cao, Y., Li, D. and Zhang, J. 2002. Progress in the research of iron limitation to marine phytoplankton. Marine Science Bulletin/ Haiyang Tongbao 21(6):83-89.

Cavender-Bares, K. K., Mann, E. L., Chisholm, S. W., Ondrusek, M. E. and Bidigare, R. R. 1999. Differential response of equatorial Pacific phytoplankton to iron fertilization. Limnology and Oceanography 44(2):237-246.

Coale, K. H., Fitzwater, S.E., Gordon, R.M., Johnson, K. S. and Barber, R. T. 1996. Control of community growth and export production by upwelled iron in the Equatorial Pacific Ocean. Nature 379: 621-624.

Coale, K. H., Johnson, K. S., Fitzwater, S. E., Gordon, R. M., Tanner, S., Chavez, F. P., Ferioli, L., Sakamoto, C., Rogers, P., Millero, F., Steinberg, P., Nightingale, P., Cooper, D., Cochlan, W. P. and Kudela, R. 1996. A massive phytoplankton bloom induced by an ecosystem-scale iron fertilization experiment in the Equatorial Pacific Ocean. Nature 383:495-501.

Coale, K. H., Steele, J. H., Turekian, K. K. and Thorpe, S. A. 2001. Iron fertilization. In: Encyclopedia of Ocean Sciences 3: 1385-1397, Academic Press, London.

Weatherhead, E. C. and Andersen, S. B. 2006. The search for signs of recovery of the ozone layer. Nature 441: 39-45.

Epstein, P., Diaz, H., Elias, S., Grabherr, G., Graham, N., Martens, W., Thompson, E. M. and Susskind, J. 1998. Biological and physical signs of climate change: focus on mosquito borne diseases. Bulletin of the American Meteorological Society 79: 409-417.

Falcam, L. 2001. Death by Warming. Honolulu Advertiser. August 12.

Fitzwater, S. E., Coale, K. H., Gordon, R. M., Johnson, K. S. and Ondrusek, M. E. 1996. Iron deficiency and phytoplankton growth in the Equatorial Pacific. Deep-Sea Research (Part II, Topical Studies in Oceanography) 43: 995-1015.

Frost, B.W. 1996. Phytoplankton bloom on iron rations. Nature 383:475-476.

Gervais F, Riebesell, U. and Gorbunov M. Y. 2002. Changes in primary productivity and chlorophyll a in response to iron fertilization in the Southern Polar Frontal Zone. Limnology and Oceanography. 47(5):1324-1335

Glantz, M. H. 1996. Currents of Change: El Nino's impact on climate and society. Cambridge University Press, Cambridge.

Goulder, Lawrence, H. and Stephen and Schneider, L. 1999. Induced Technological Change and the Attractiveness of CO2 Emissions Abatement Policies. Resource and Energy Economics 21:211-253.

Grabherr, G., Gottfried, M. and Pauli, H. 1994. Climate effects on mountain plants. Nature 369: 448.

Gullison R. E. et al. 2007. Tropical forests and climate policy Science 316: 985-986. IPCC 2007. Climate Change 2007: The Physical Science Basis: Summary for Policymakers. http://www.ipcc.ch/pdf/asessment-report/ar4/wg1/ar4-wg1-spm.pdf.

Hajat, S., Ebi, K. L., Edwards, S., Haines, A., Kovats, S. and Menne, B. 2003. Review of the human health consequences of flooding in Europe and other industrialized civilizations. Appl. Environ. Sci. Public Health 1:13-21.

Hamnett, M. P. 1998. The Pacific ENSO Applications Center and the 1997-1998 El Nino. December 1998.

Hoel, M. and K. Schneider. 1997. Incentives to Participate in an International Environmental Agreement. Environment and Resource Economics 9:153-170.

IPCC, 1998. The Regional Impacts of Climate Change: An Assessment of Vulnerability. (Eds. R. T. Watson, M. C. Zinyowera and R. H. Moss), Cambridge University Press, Cambridge,.

Jacoby, Henry D., and Denny Ellerman A. 2004. The Safety Valve and Climate Policy. Energy Policy 32 (4):481-491.

Kiehl, J. T. and Trenberth, K. 1997. Earth's annual mean global energy budget, Bulletin of the American Meteorological Society 78: 197–208.

Kolstad, Charles D. 1996. Learning and stock effects in environmental regulation: the case of greenhouse gas emissions. Journal of Environmental Economics and Management 31 (1):1-18.

Lal, M, Harasawa, H. and Takahashi K. 2002. Future climate change and its impacts over small island states. Climate Res. 19:179-192.

Lean, Geoffrey. 2007. Are cows killing the planet? Daily Mail. July 11th Issue.

Manne, Alan S. and R. Richels. 2004. The impacts of learning-by-doing on the timing and costs of CO2 abatement. Energy Economics 26 (4):603-619.

Mansur, Erin, Robert Mendelsohn and Wendy Morrison. 2005. A Discrete-Continuous Choice Model of Climate Change Impacts on Energy. New Haven, CT: Yale School of Forestry and Environmental Studies.

Martin, J. H., Gordon R.M. and Fitzwater S.E. 1990. Iron in Antarctic waters. Nature 345:156-158.

Martin, J. H. 1990. Glacial-interglacial CO_2 change: The iron hypothesis. Paleoceanography 5(1):1-13.

Martin, J. H., Gordon, R.M. and Fitzwater, S.E. 1990. Iron in Antarctic waters. Nature 345:156-158.

McFarland, J. R., Reilly, J. M. and H. J. Herzog. 2004. Representing energy technologies in top-down economic models using bottom-up information. Energy Economics 26 (4):685-707.

Meehl, G. A. 1996. Vulnerability of freshwater resources to climate change in the tropical Pacific region. Water Air Soil Pollut. 92:203-213.

Meehl, G. A. 1997. Pacific region climate change. Ocean Coastal Manag. 37:137-147.

Mendelsohn, Robert. 2003. Assessing the market damages from climate change. In Global Climate Change: The Science,

Economics, and Politics. Cheltenham, UK.

Newman, P. A., Kawa, S. R. and Nash, E. R. 2004. On the size of the Antarctic ozone hole? Geophysical Research Letters 31: L12814.

Nicholls, R. J. and Leatherman, S. P. 1996. Global sea-level rise. In: As Climate Changes: International Impacts and Implications (Eds. Strzepek, K. M and Smith, J. B.), pp. 92-123, Cambridge University Press, Cambridge.

Nurse, L. A., Sem, G., Hay, J. E., Suarez, A.G, Wong, P. P. and Briguglio L. 2001. Small island states. In: Climate Change 2001. Impacts, Adaptation and Vulnerability. Contribution of Working Group II to the Third Assessment Report of the Intergovernmental Panel on Climate Change (Eds. McCarthy, J. J., Canziani, O. F., Leary, N. A., Dokken, D .J. and White K. S.), pp. 844-875, Cambridge University Press, Cambridge.

Oerlemans, J. 2005. Extracting a Climate Signal from 169 Glacier Records. Science 308: 675–677.

Parmesan, C. 1996. Climate and species range. Nature 382: 765-766.

Parmesan, C. et al. 1999. Poleward shifts in geographical ranges of butterfly species associated with regional warming. Nature 399: 579-583.

Smith, R. et al. 1992. Ozone depletion: Ultraviolet radiation and phytoplankton biolgy in antarctic waters Science 255: 952.

R. P. Wayne 1991. Chemistry of the Atmospheres 2nd edition, Oxford.

Revkin, Andrew C. 2005. Rice in gases unmatched by a history in ancient Ice. The New York Times.

Roan, Sharon 1990. Ozone Crisis, the 15 Year Evolution of a Sudden Global Emergency. Wiley.

Ropelewski, C. F. and Halpert, M. S. 1987. Global and regional scale precipitation patterns associated with the El Nino/Southern Oscillation. Monthly Weather Rev 115:1606-1625.

Royer, J. F., Chauvin, F., Timbal, B., Araspin, P. and Grimal, D. 1998. A GCM study of the impact of greenhouse gas increase on the frequency of occurrence of tropical cyclones. Clim Change 38:307-343.

Ruddiman, W. F. 2003. The anthropogenic greenhouse era began thousands of years ago. Climate Change 61 (3): 261-293.

Ruddiman, W. F. 2005. Earth's climate past and future. Princeton University Press, New York.

Ruttimann, Jacqueline 2006. Doomsday food store takes pole position. Nature 441: 912–913.

Santilli M. et al. 2005 Tropical deforestation and the Kyoto Protocol: an editorial essay Clim. Change *71: 267-276.*

Schmidt, C.W. 2005. Keeping afloat: a strategy for small island nations. Environ Health Perspect 113:A606-A609.

Smith, Lewis. 2007. How to stop cows burping is the new field work on climate change. The Times July 10th Issue.

Spennemann, D. H. R. and Marschner, I.C. 1995. The association between El Nino/Southern Oscillation events and typhoons in the Marshall Islands. Disasters 19:194-197.

Thomas, C.D. and Lennon, J.J. 1999. Birds extend their ranges northwards. Nature 399: 213.

Watson, R. T, Zinyowera, M. C., Moss, R. H. and Dokken, D .J. Eds. 1996. Impacts, Adaptations and Mitigation of Climate Change: Scientific-Technical Analyses. Contribution of Working Group II to the Second Assessment Report of the Intergovernmental Panel on Climate Change. Cambridge, UK: Cambridge University Press.

WHO. 2000. Climate Variability and Change and their Health Effects in Pacific Island Countries. WHO/SDE/OEH/01.1, World Health Organization, Geneva.

WHO. 2002. World Health Report 2002: Reducing Risks, Promoting Healthy Life. World Health Organization, Geneva.

WHO. 2003. Synthesis Workshop on Climate Change and Health. 1-4 December 2003, Republic of the Maldives. World Health Organization, Geneva.

William E. F. 2007. Seeds of Destruction, Montreal. Global Research.

"Doomsday' seed arrive in Norway", BBC News Online, 31st January 2008.

"Doomsday' seed vault opens in Arctic", boston.com/Associated Press, 26th February 2008.

http://www.timesonline.co.uk/tol/news/uk/science/article2051364.ece

http://nzsap.org.nz/proc/2004/ab04039.html

http://www.eatwild.com/cla.html

http://www.epa.gov/agriculture/ag101/dairyglossary.html

http://news.bbc.co.uk/2/hi/uk_news/wales/mid_/6288012.stm

http://mbreiding.us/ert/Arizona/desbiome/nitrofix.htm

http://www.agr.state.nc.us/cyber/kidswrld/plant/nutrient.htm

http://www.sustainabletable.org/intro/dictionary/

http://www.ansci.cornell.edu/plants/toxicagents/tannin/pos_effects.html

http://www.agu.org/revgeophys/chisho00/chisho00.html

http://tracer.env.uea.ac.uk/soiree/

http://www.mbari.org/expeditions/SOFeX2002/

http://www.planktos.com/main.htm

http://www.niwa.cri.nz/pubs/wa/10-4/newsforum

http://www.ems.psu.edu/~fraser/Bad/BadGreenhouse.html

http://www.croptrust.org/main/donors.php.

INDEX

D

E

F

G

H

I

J

K

L

M

U

W

Y